Fire Resistance of Aluminum and Aluminum Alloys

& Measuring the Effects of Fire Exposure on the Properties of Aluminum Alloys

J. Gilbert Kaufman

ASM INTERNATIONAL

ASM International
Materials Park, OH 44073-0002
asminternational.org

First printing, June 2016

Comments, criticisms, and suggestions are invited, and should be forwarded to ASM International.

Prepared under the direction of the ASM International Technical Book Committee (2015–2016), Y. Zayna Connor, Chair.

ASM International staff who worked on this project include Scott Henry, Director, Content and Knowledge-Based Solutions; Karen Marken, Senior Managing Editor; Sue Sellers, Content Development and Business Coordinator; Madrid Tramble, Manager of Production; Diane Whitelaw, Production Coordinator; and Kelly Sukol, Project Coordinator.

Library of Congress Control Number: 2015959457
ISBN-13: 978-1-62708-061-1
EISBN: 978-1-62708-107-8
SAN: 204-7586

ASM International®
Materials Park, OH 44073-0002
asminternational.org

Printed in the United States of America

Contents

Preface

Aluminum melts at approximately 660 °C (1220 °F), lower than most common structural metals such as iron and steel. Because of this, its behavior in fires can lead to confusion about its performance. Work was expended on this publication to document facts about the fire resistance of aluminum and aluminum alloys, and to enable engineers and designers to take account of aluminum's characteristic high resistance to burning while recognizing its relatively low melting point. The information includes facts with corresponding references; speculation and subjectivity are excluded.

Other publications have provided very useful technical data and guidance concerning some aspects of dealing with the characteristics of aluminum alloys with respect to fire exposure, but none has provided the full scope of coverage contained here.

The author gratefully acknowledges the support of the Aluminum Association, Inc. for access to its publications and photographs, the support of Alcoa, Inc. for access to previously unpublished data for aluminum alloys included herein, and to ASM International.

About the Author

John Gilbert (Gil) Kaufman has a background of over 50 years in the aluminum and materials information industries and remains an active consultant in both areas. In 1997, he retired as Vice President, Technology for the Aluminum Association, Inc., headquartered then in Washington, D.C. (now in Arlington, Va.), and is currently president of his own consultancy, Kaufman Associates.

Earlier in his career, Kaufman spent 26 years with the Aluminum Company of America, where he managed engineering properties and fabricating metallurgical research at Alcoa Laboratories. Many of the data presented in this volume were generated over the period when the author was active in and/or managing Alcoa Laboratories engineering properties research.

Kaufman subsequently spent five years with ARCO Metals, where he was Director of R&D and, later, Vice President, Research & Engineering. Kaufman also served for nine years as President and CEO of the National Materials Property Data Network where, working with STN International and Chemical Abstracts Service, he established a worldwide online network of more than 25 numeric materials properties databases.

Gil is a Fellow and Honorary Member of ASTM and a Fellow and Life Member of ASM International. He is a licensed professional engineer in Delaware. Gil has published more than 140 articles and six books on aluminum alloys and materials data systems.

Introduction

The natural physical characteristics of aluminum and its alloys are such that they do not burn under normal atmospheric conditions nor do they contribute to flame spread or act as a fire accelerant. The data supporting these statements are provided in detail in the following chapters. Other organizations have addressed various aspects of this subject quite well (Ref 1–6), but none has addressed the whole scope of relevant material as attempted here.

The performance of aluminum alloys is excellent in many applications requiring exposure to relatively high temperatures, including (a) personal and commercial vehicles of many kinds, (b) marine applications such as fast ferry hulls and oil drilling rigs where superior corrosion resistance is critical, and (c) a variety of structural applications such as buildings, bridges, and pressure vessels. More details about the importance of aluminum's high resistance to burning in some of these applications is discussed.

Because aluminum melts at a temperature of approximately 655 °C, or 1200 °F (Ref 7, 8), lower than most common structural metals such as iron and steel, its behavior in some structural situations can lead to misunderstandings about its performance in fires. For example, when vehicles such as cars, trucks, or ships with aluminum components are caught in an engulfing fire, the aluminum components may be reported to have burned because they appear to combust and burn away. In fact, the aluminum components melt and run off, giving the appearance of being consumed in the fire. Aluminum or aluminum alloy components do not burn or contribute to the combustion.

Misunderstandings about aluminum behavior in fires has occurred in the past. For example, during the Falklands War in 1982, it was widely reported that the British warship HMS *Sheffield* was destroyed by Argentine rockets because the ship was made largely of aluminum and had combusted; in fact, the *Sheffield* was made entirely of steel and its destruction had nothing to do with aluminum (Ref 9).

It is the purpose of this book to describe the facts regarding the behavior of aluminum at very high temperatures, including those as high as or higher than necessary to cause it to melt, and to characterize its behavior in a wide range of applications where high-temperature performance is important. The behavior described is based entirely on documented test data, primarily the results of tests made in accordance with ASTM Standard Methods (Ref 10–12) and British Standards (Ref 13–17).

Consideration is also given to situations where aluminum alloys have been exposed to fire but not melted and there is a subsequent need to estimate the residual strength of the exposed members. Although aluminum alloys lose strength when exposed to temperatures above approximately 100 °C, or 212 °F (Ref 18), they do not deform until temperatures near 500 to 600 °C (800 to 900 °F) are reached. Engineers must be able to determine whether aluminum alloy components that have had exposure to fire are able to continue to function satisfactorily or if performance has been compromised. By using nondestructive tests such as hardness and electrical conductivity measurements, it is possible to estimate with considerable accuracy the retained mechanical strength. Tools are included herein to guide estimates of this type.

There are, of course, applications where the high-temperature exposure is too great for aluminum to be used due to its low melting temperature. These are also documented, and guidance is provided for decisions on whether or not to use aluminum alloys in new applications.

The facts concerning the fire resistance of aluminum are:

- The physical properties of aluminum, notably high thermal conductivity, specific heat, and reflectance and its low emissivity, provide resistance to structures against temperature rise comparable or superior to that provided by steels in the early stages or in a non-engulfing fire.
- Even when temperatures do increase to intense incendiary levels, aluminum does not burn in air nor will it support combustion. When tested in accordance with ASTM or British standards, aluminum provides the highest ratings for resisting flame spread because it is not easily ignitable under atmospheric conditions and does not support flame spread.
- When necessary, the structural integrity of aluminum alloy structures can be protected against fire by practical and commercial fireproofing technology such as lightweight vermiculite concrete, similar to that used to fire protect steels, or Rockwool or gypsum sheeting.
- Aluminum is nonsparking in all environments and with all materials, with one known exception: when bare (unpainted or uncoated) aluminum is struck by or strikes rusty ferrous metals, sparks may result.
- Under conditions where it is likely or possible that aluminum may be struck by rusty ferrous metals, protective coatings such as paint are recommended to avoid any possibility of sparking.

REFERENCES

1. J.A. Purkiss and L.-Y. Li, *Fire Safety Engineering Design of Structures,* 3rd ed., CRC Press, New York, 2013
2. B. Faggiano, G. De Matteis, R. Landolfo, and F.M. Mazzolani, Behaviour of Aluminium Structures Under Fire, *J. Civ. Eng. Manag.,* Vol X (No. 3), 2002, p 183–190
3. M.J. Bayley, The Fire Protection of Aluminium in Offshore Structures, *Proceedings of the International Mechanical Engineering (IMechE) Conference on Materials and Design against Fire,* Mechanical Engineering Publications, London, 1992, p 113–120
4. S. Lundberg, "Material Aspects of Fire Design," TALAT Lecture 2502, European Aluminium Association, 1994
5. "Fire Resistance and Flame Spread Performance of Aluminum and Aluminum Alloys," Standard AA FRFS, 2nd ed., The Aluminum Association, Washington, D.C., July 2002
6. Fire Resistance of Aluminum, *Aluminum and the Sea,* Alcan Aluminium Company, 2013
7. *Aluminum Standards and Data 2013,* The Aluminum Association, Arlington, VA, 2013
8. *Aluminum Standards and Data 2013 Metric SI,* The Aluminum Association, Arlington, VA, 2013
9. "The Falklands Campaign: The Lessons," presented to Parliament by the Secretary of Defence by Command of Her Majesty, Dec 1982
10. "Standard Methods of Fire Tests of Roof Coverings," ASTM E108, *Annual Book of ASTM Standards,* ASTM (updated annually)
11. "Standard Test Methods for Fire Tests of Building Construction and Materials," ASTM E119, Part 04.07, *Annual Book of ASTM Standards,* ASTM (updated annually)
12. "Standard Test Method for Behavior of Materials in a Vertical Tube Furnace at 750 °C," ASTM E136, *Annual Book of ASTM Standards,* ASTM (updated annually)
13. "Classification of Materials for Fire Resistance, Part 3: External Fire Exposure Roof Test," British Standard 476, The British Standards Institution, 1975
14. "Classification of Materials for Fire Resistance, Part 4: Non-combustibility Test for Materials," British Standard 476, The British Standards Institution, 1970
15. "Classification of Materials for Fire Resistance, Part 5: Ignitability of Building Materials" (now replaced by Part 4), British Standard 476, The British Standards Institution
16. "Classification of Materials for Fire Resistance, Part 6: Fire Combustibility of Coated Systems" (now obsolete), British Standard 476, The British Standards Institution
17. "Classification of Materials for Fire Resistance, Part 23: Methods for Determination of the Contribution of Components to the Fire

Resistance of a Structure," British Standard 476, The British Standards Institution

18. J.G. Kaufman, *Properties of Aluminum Alloys: Tensile, Creep and Fatigue Data at High and Low Temperatures,* ASM International, Materials Park, OH, 1999

CHAPTER 1

Properties and Characteristics of Aluminum and Aluminum Alloys

1.1 Melting of Aluminum and its Alloys

Unalloyed aluminum melts at a temperature of approximately 655 °C (1215 °F); it boils at approximately 2425 °C (4400 °F) (Ref 1.1, 1.2). Alloys of aluminum do not melt at a fixed temperature but rather over a range of temperatures dependent on their composition. For example, alloy 5456, with approximately 5% Mg alloying constituent, has a melting range of 570 to 640 °C (1055 to 1180 °F) (Ref 1.1, 1.2). Melting begins at the lower end of the range and is completed at the higher end. The melting ranges for many commonly used aluminum alloys are provided in Table 1.1.

Table 1.1 Melting ranges of some representative aluminum alloys

Alloy	Melting range °C	Melting range °F	Alloy	Melting range °C	Melting range °F
1100	640–655	1190–1215	5052	605–650	1125–1200
2024	500–635	935–1180	6061	580–650	1080–1205
3003	640–655	1190–1210	7075	475–635	890–1175

Source: Ref 1.1, 1.2

Aluminum and aluminum alloys are melted and remelted regularly as needed for the casting of ingots or billets for subsequent fabricating procedures such as rolling, extruding, drawing, or forging and also for recycling. Aluminum does not ignite or catch fire as it is being melted nor does it emit smoke or toxic gases.

1.2 Mechanical Properties of Aluminum Alloys at High Temperatures

The properties of aluminum alloys are compromised at elevated temperatures well before the metal reaches its melting temperature (Ref 1.3). For most of the alloys, strengths after significant times at temperatures above 150 to 200 °C (300 to 400 °F) are lower than those at room temperature, and the amount of the strength reduction may increase with both increasing temperature and/or increasing time at an elevated temperature. As a result, most aluminum alloys are not usually recommended for long-time service at or above these temperatures, but they are widely used in the temperature range from room temperature up to 150 to 200 °C. Certain alloys specifically designed to maximize high-temperature resistance, such as those in the 2*xxx* aluminum-copper series, are usually chosen for applications in the higher end of this range.

Tables illustrating the high-temperature tensile properties of representative commercial aluminum alloys are included in Appendix 1.

1.3 Physical Properties of Aluminum Alloys

Several of the physical properties of aluminum and its alloys provide some protection when the alloys are near a fire in an adjacent structure and also lessen their increase in temperature in the early stages of a more immediate fire. Those physical properties include (Ref 1.4–1.7):

- The specific heat capacity of aluminum alloys (816 to 1050 J/kg · K, or 0.195 to 0.258 Btu/lb · °F), which is approximately twice that of steel (377 to 502 J/kg · K, or 0.090 to 0.120 Btu/lb · °F) (Ref 1.7). This means that it takes twice as much heat energy to raise the temperature of aluminum one degree as compared to a similar mass of steel. So in any fire, aluminum members would be relatively slower to heat. This advantage is retained as temperature increases, because the specific heat of aluminum alloys increases with temperature to the melting point (Ref 1.4).
- The thermal conductivity of aluminum and its alloys, which is 88 to 251 W/m · K, or 51 to 164 Btu (h · ft · °F), and increases with increase in temperature (Ref 1.4). This is several times the value for steels (11 to 63 W/m· K, or 6 to 37 Btu [h · ft · °F]) (Ref 1.7). Thus, heat from a localized source will be distributed along an aluminum structure in a

much more efficient manner, enabling it to be radiated off and minimizing hot spots. Also, if the structure is sufficiently massive, the aluminum can act as a heat sink to slow the rate of increase of temperature in the early stages of a fire, increasing the period of serviceability. This might make the difference in prolonging structural endurance in a fire and allowing time to evacuate a burning structure.

- The reflectivity of aluminum, which is very high—80 to 90% of incident radiation, many times that of bare steel, and reportedly 17 to 19 times greater than the usual painted steel structures (Ref 1.7). It remains very high, even at high temperatures and even for old and oxidized surfaces. Thus for bare aluminum or aluminum alloys, this high reflectivity also contributes to a slower rise in temperature and longer serviceability than for most structural steels during the early stages of a fire. Reflectivity is decreased if the aluminum surfaces are painted or become coated with soot.
- The emissivity of aluminum alloys (0.02 to 0.10 ε for most structural aluminum alloys), which is lower than that of carbon steels (0.10 to 0.80 ε) and stainless steels (0.27 ε) (Ref 1.7). This also contributes to the ability of aluminum alloys to heat up more slowly than steels in the early stages of a fire, allowing more time for occupants to escape the fire. While emissivity varies greatly depending on surface quality and cleanliness, steel members may heat up approximately four times faster than comparable aluminum alloy members in a non-engulfing fire (Ref 1.8).

As noted, these physical properties are most important if the aluminum components of the structure are nearby or adjacent to the main fire in another structure, but they may also be helpful in the very early stages of a serious conflagration in the immediate structure. If the aluminum members become heavily coated with soot, the advantages offered by the physical properties of the original components are diminished or nonexistent.

The physical properties of several typical aluminum alloys and a widely used structural steel are illustrated in Table 1.2 (Ref 1.2, 1.7). More com-

Table 1.2 Physical properties of representative aluminum alloys and steel

Material	Melting range, °C	Boiling point, °C	Melting heat, $kJ \cdot kg^{-1}$	Specific heat, $J/kg^{-1} \cdot K^{-1}$	Thermal conductivity	Emissivity, ε	Coefficient of thermal expansion, $10^{-6} \cdot K^{-1}$, 20–100 °C
1050-O	645–658	2425	390	900	229	0.02–0.10	23.5
5083-O	574–638	2425	390	900	117	0.02–0.40	24.2
6005A-T5	605–655	2425	390	940	188	0.02–0.40	23.6
ASTM E24 steel	1400–1530	2860	250	420	54	0.10–0.80	13.5

plete tables of the physical properties of aluminum alloys are included in Appendix 2.

1.4 Resistance to Burning in Normal Atmospheric Conditions

As illustrated in the tests described subsequently, solid bulk aluminum will not burn and has never been observed to burn in air. Similarly, molten aluminum has not been observed to burn in air. In neither situation does aluminum give off smoke or any hazardous fumes. The natural oxide coating on solid aluminum forms very rapidly and inhibits reaction of the underlying solid aluminum to air, thereby contributing to its high resistance to burning.

Like finely divided metallic powders of most metals, aluminum powder is very flammable and is hazardous to handle (Ref 1.8); it is used to make explosives. In a fire, this behavior is entirely different from that of solid or molten aluminum. Even thin foils of aluminum are impossible to get to burn rather than melt.

Aluminum has been thoroughly evaluated for structures where fire may be encountered and is given the highest rating for such applications by ASTM Standards (Ref 1.9–1.11), British Standard 476 (Ref 1.12–1.16), European Communities Directives on Construction Products (Ref 1.17), and various U.S. building codes (e.g., Ref 1.18).

1.4.1 ASTM Standards (Ref 1.9–1.11)

ASTM Standard E108. Fire tests to determine combustibility of aluminum structural components of aluminum roofs and dome structures were made for TEMCOR Co. by United States Testing Company in accordance with ASTM Standard E108, "Standard Methods of Fire Tests of Roof Coverings." This test method was comparable to the fire test standards of Los Angeles Building Code 5702.01 (Ref 1.18), Underwriters' Laboratories Standard UL 790 (Ref 1.19), and National Fire Protection Association Standard 256 (Ref 1.20). Measurements were made of dimensional stability, weight loss, and appearance changes of pieces of the space frame truss. Aluminum sample panels, 1.397 mm (0.055 in.) thick, were exposed to temperatures up to 825 °C (~1500 °F) for up to 10 min. There was some melting of the thin roof panels but no combustion, and, in fact, there were no dimensional changes of the space frame components observed. These tests and the results were described in two United States Testing Company Reports dated August 6 and 7, 1985 (Ref 1.21, 1.22).

ASTM Standard E136. A number of different aluminum alloys were tested by Signet Testing Laboratories in conformance with ASTM Standard E136-65, "Combustibility of Materials in a Vertical Tube Furnace." The alloys were tested at 750 °C (1380 °F) (Ref 1.5) for Kaiser Aluminum

& Chemical Company in the period from 1968 to 1972. Reports were issued by Signet dated September 30, 1968, covering alloys 3004 and 8112, and May 17, 1972, covering alloys 3003, 3105, and 5005 (Ref 1.23, 1.24). All alloys were rated "noncombustible."

Copies of representative reports documenting evaluations of the fire resistance of aluminum alloys and aluminum structures are contained in Appendix 3.

1.4.2 British Standards (Ref 1.12–1.16)

Part 4 and now-obsolete Part 5 of BS 476 provided for tests for noncombustibility and ignitability, respectively, of structural materials (Ref 1.13, 1.14). Aluminum alloy test pieces, 40 mm (1.6 in.) in width and breadth and 50 mm (2 in.) in height, were exposed in a furnace to a stabilized temperature of 750 °C (1380 °F) for a period of more than 10 min. During this exposure, continuous observations were made on (a) whether the temperature in the furnace increased by 50 °C (122 °F) or more, which would indicate the material contributed to an increase in temperature, and (b) whether or not there was any period of flaming in the furnace for 10 s or more, which would indicate ignition. Aluminum alloys were not observed to ignite, flame, or contribute in any way to the temperature rise in the furnace. They were rated P for "not easily ignitable."

Aluminum was also tested in accordance with British Standard 476, Part 3, for flame spread and fire penetration of roof structures (Ref 1.12). In this test, aluminum alloy roofing structure samples at least 1.5 by 1.2 m in thickness were exposed to test flames of luminous coal gas or natural gas 200 to 250 mm long. External surfaces of aluminum demonstrated the highest resistance to both fire penetration and flame spread and were classified as AA. For inner surfaces, aluminum demonstrated very high resistance to flame spread and was classified as 0, the highest rating for that type of assembly (Ref 1.21, 1.23, 1.24).

Part 6 of BS 476 covers fire propagation performance for coated systems (Ref 1.15). Because of its hard oxide coating and excellent corrosion resistance with the need for only thin protective coatings, aluminum consistently achieves high ratings in this situation as well.

1.4.3 National Standard of Canada CAN4-S114-M80 (Ref 1.25)

In 1982, noncombustibility tests were run on aluminum alloy 6063 by the National Research Council of Canada in accordance with their National Standard of Canada CAN4-S114-M80. The tests were run in triplicate, with three specimens, 3.8 by 3.8 by 5.0 cm (1.5 by 1.5 by 2.0 in), held in a furnace stabilized to 750 °C (1380 °F) for a minimum of 15 min while being visually examined for flaming or smoking and subsequently weighed for weight loss. The conclusions from the tests were that aluminum "met the requirements for non-combustibility according to CAN4

S114-M80 since (a) maximum temperature rise was zero, (b) sample did not flame during the test, and (c) maximum weight loss did not exceed 20 percent." The results were reported in NRC Report E-11-67, dated June 9, 1982, written by R.C. Monette and submitted by T. Harmathy (Ref 1.26).

1.4.4 Uniform Building Code (Ref 1.27)

Alloys 6061-T6 and 6063-T5 were tested in accordance with the requirements for incombustible materials of the Uniform Building Code published by the International Conference of Building Officials (Ref 1.27). In these tests, three pieces of structural extrusions of each alloy were subjected to temperatures of 650 to 655 °C (1205 to 1210 °F) for a period of 5 min with no observed ignition or flaming. They were all noted to conform to the requirements for an "incombustible" rating.

1.5 Burning in Pure Oxygen

Rapid oxidation of aluminum and other metals, including steel, has been reported in several laboratory investigations using a 100% oxygen environment (Ref 1.28–1.33). In these studies, solid aluminum was forced to oxidize rapidly when an oxygen-gas flame was trained directly on the aluminum specimen, melting the surface. Even then, rapid oxidation or burning occurred only after the oxide layer was mechanically removed. When the oxygen stream was removed, the reaction immediately stopped.

In a review article (Ref 1.28), the generalization was stated that "all metals, with the possible exception of gold and platinum, can be expected to ignite in oxygen at some elevated temperature." Ignition-sensitive alloy systems were defined as alloys of titanium, zirconium, thorium, uranium, lead, tin, and magnesium. The article goes on to say that alloy systems rated to be relatively insensitive to ignition in an oxygen environment include austenitic stainless steels, nickel alloys, cobalt alloys, copper alloys, and silver alloys. A third group of alloys was described as intermediate between the sensitive and insensitive groups; that group includes aluminum alloys, carbon steels, low-alloy high-strength steels, and 400-series stainless steels.

It is clear that a 100% oxygen environment is required to get any rapid oxidation or ignition of aluminum and aluminum alloys as well as steels, and that any combustion stops immediately if the supply of pure oxygen is stopped.

1.6 Resistance to and Protection from Thermic Sparking

Accidents in the mining industry during the 1950s were attributed to the thermic reaction of aluminum striking or being struck by rusty steel. Uppal

(Ref 1.34) indicates that perhaps the greatest fear of offshore engineers in using aluminum components is the possibility of an explosion resulting from an exothermic reaction between rusty steel and aluminum creating a spark when the piece of aluminum strikes a steel component; this is referred to as thermite sparking.

Though relatively rare, these events spurred on a great deal of research by the aluminum industry, and the nature and methods for protection against such thermic reactions are now well understood (Ref 1.34–1.36).

Thermic sparking occurs when a blow of aluminum against rusty iron or steel results in a transfer of oxygen between intimately mixed aluminum and rust (iron oxide) particles. Explosions may result if the thermic sparking occurs in the presence of an ignitable environment.

However, it is important to note that thermic sparking requires a very specific set of pre-conditions to exist simultaneously at the time of contact, and these conditions are rarely met. Thermic sparking does not occur when aluminum is struck in a normal ambient atmosphere by other aluminum, nor with any other material, including non-rusty iron and steel. So overall, the likelihood of thermic sparking even under hazardous conditions is considered low, and it is essentially nonexistent under normal atmospheric conditions.

In those situations where there is some concern that aluminum might be directly in contact with rusty iron or steel in the presence of an ignitable environment of any kind, it is recommended that the aluminum surfaces be painted and the paint maintained in good condition.

Despite the original mining accidents that prompted so much study of thermic reactions, aluminum is now widely used and recommended for mining applications and has been for many years. For more detail on such applications and on the low risk of reactions in mining situations, the reader is referred to Ref 1.35.

REFERENCES

1.1 *Aluminum Standards and Data 2013,* The Aluminum Association, Arlington, VA, 2013

1.2 *Aluminum Standards and Data 2013 Metric SI,* The Aluminum Association, Arlington, VA, 2013

1.3 J.G. Kaufman, *Properties of Aluminum Alloys: Tensile, Creep and Fatigue Data at High and Low Temperatures,* ASM International, Materials Park, OH, 1999

1.4 S. Lundberg, "Material Aspects of Fire Design," TALAT Lecture 2502, European Aluminium Association, 1994

1.5 "Fire Resistance and Flame Spread Performance of Aluminum and Aluminum Alloys," Standard AA FRFS, 2nd ed., The Aluminum Association, Washington, DC, July 2002

1.6 Fire Resistance of Aluminum, *Aluminum and the Sea,* Alcan Aluminium Company, 2013

1.7 F. Cverna, Ed., *ASM Ready Reference: Thermal Properties of Metals,* ASM International, 2002
1.8 R. Pape and F. Schmidt, Combustibility Analysis of Metals, *Adv. Mater. Process.,* Dec 2009, p 41–44
1.9 "Standard Methods of Fire Tests of Roof Coverings," ASTM E108, *Annual Book of ASTM Standards,* ASTM (updated annually)
1.10 "Standard Test Methods for Fire Tests of Building Construction and Materials," ASTM E119, Part 04.07, *Annual Book of ASTM Standards,* ASTM (updated annually)
1.11 "Standard Test Method for Behavior of Materials in a Vertical Tube Furnace at 750 °C," ASTM E136, *Annual Book of ASTM Standards,* ASTM (updated annually)
1.12 "Classification of Materials for Fire Resistance, Part 3: External Fire Exposure Roof Test," British Standard 476, The British Standards Institution, 1975
1.13 "Classification of Materials for Fire Resistance, Part 4: Non-combustibility Test for Materials," British Standard 476, The British Standards Institution, 1970
1.14 "Classification of Materials for Fire Resistance, Part 5: Ignitability of Building Materials" (now replaced by Part 4), British Standard 476, The British Standards Institution
1.15 "Classification of Materials for Fire Resistance, Part 6: Fire Combustibility of Coated Systems" (now obsolete), British Standard 476, The British Standards Institution
1.16 "Classification of Materials for Fire Resistance, Part 23: Methods for Determination of the Contribution of Components to the Fire Resistance of a Structure," British Standard 476, The British Standards Institution
1.17 "94/611/EC: Commission Decision of 9 September 1994 Implementing Article 20 of Directive 89/106/EEC on Construction Products," European Economic Community, Sept 9, 1994
1.18 "Combustible Material," Los Angeles City Municipal Code, Los Angeles, CA, paragraph 5702.01
1.19 "Standard Fire Test Method for Roof Coatings," UL 790, Underwriters Laboratories, Northbrook, IL
1.20 "Standard Methods of Fire Tests of Roof Coverings," NFPA 256, National Fire Protection Association, Quincy, MA, 2003
1.21 "Structural Materials Employed in the TEMCOR Aluminum Dome," Report, United States Testing Company, Inc., Los Angeles, CA, Aug 6, 1985
1.22 "Roof Fire Test Evaluation TEMCOR Aluminum Dome Panel," Report, United States Testing Company, Inc., Los Angeles, CA, Aug 7, 1985
1.23 Lab Report No. 4342, Signet Testing Laboratories, Hayward, CA, Sept 30, 1968, prepared for Kaiser Aluminum & Chemical Co., re-

porting tests dated Aug 23, 1968 (Alloy 3004) and Sept 17, 1968 (Alloy 8112)

1.24 Lab Report 10263, Signet Testing Laboratories, Hayward, CA, May 17, 1972, prepared for Kaiser Aluminum & Chemical Co., reporting tests dated May 5, 1972 (Alloys 3003, 3105, 5005)

1.25 "Standard Method of Test for Determination of Non-Combustibility in Building Materials," National Standard of Canada CAN4-S114-M80, Underwriters' Laboratories of Canada, Dec 1980

1.26 R.C. Monette and T. Harmathy, "Non-Combustibility Test in Accordance with CAN4-S114-M80," Canadian National Research Council Report No. E-11-67, June 9, 1982

1.27 Uniform Building Code Standard No. 4-1-6, Section 410, Vol I & III, 1961 ed., International Conference of Building Officials

1.28 "Ignition of Metals in Oxygen," DMIC Report 224, Feb 1, 1961

1.29 A.H. Tench, H.M. Roder, and A.F. Clark, "Combustion of Metals in Oxygen, Phase II: Bulk Burning Experiments," NBSIR Report 73-345, National Bureau of Standards, Boulder, CO, Dec 1973

1.30 A. Lapin, "Oxygen Compatibility of Materials," presented at the International Institute of Refrigeration, Nov 1973

1.31 A. Macek, Fundamentals of Combustion of Single Aluminum and Beryllium Particles, *Symposium (International) on Combustion,* Vol 11 (No. 1), 1967, p 203–217

1.32 A.F. Clark and J.G. Hust, A Review of the Compatibility of Structural Metals with Oxygen, *AIAA J.,* Vol 12 (No. 4), 1974, p 441–454

1.33 D.C. Kuebl, "Ignition and Combustion of Aluminum and Beryllium," presented at the 2nd Aerospace Sciences Meeting, New York, NY, Jan 1965

1.34 N. Uppal, The Structural Use of Aluminium with Particular Reference to the Offshore Industry, *Proceedings of Alumitech'97,* May, 1997

1.35 J.T. Hurd, "Thermite Sparking and the Use of Aluminium Underground in Mining Operations," Hulett Aluminum Report No. H 90/02 CT, Hulett Aluminum Limited, February 2, 1990

1.36 "Aluminum Design Guide," Chapter 1.4.6 Fire Protection of Aluminum & Chapter 4.4 Fire Performance of Aluminium Wimpey Offshore (London) & Alcan Offshore (Gerrard Cross, UK), 1990

CHAPTER 2

Fire Protection of Aluminum Structures

A COMPREHENSIVE OVERVIEW of the variety of insulating materials that may be used to protect aluminum alloy structures in fire are presented in Section 2502-02 of TALAT Lecture 2502 (Ref 2.1). Specific guidance on some of these options is presented subsequently. All dimensions in this chapter are presented in the engineering units in which they were generated.

2.1 Vermiculite Encasement for Fire Protection

Important structural members of steel in buildings are routinely protected with various types of lightweight concrete casings, and this is also a logical means of protecting aluminum structures.

In 1962, the Aluminum Company of America evaluated procedures for fire-protecting an aluminum building structure with lightweight vermiculite concrete and published the results in the March 1963 issue of *Civil Engineering* magazine (Ref 2.2). The complete publication is reproduced in Appendix 4.

In those tests, which were conducted in accordance with ASTM Standard E119-61 (Ref 2.3–2.5) at Underwriters Laboratories outside Chicago, Ill., two extruded 2014-T6 wide-flange columns were used. The columns were 2.74 m long and 20.32 cm deep (20.32 × 20.32 × 0.732 cm, 4.86 kg per 30.48 cm of length), or 9 ft long and 8 in. deep ("8 WF 10.72": 8 × 8 × 0.288 in., 10.72 lb per ft of length). They were mounted on 2014-T6 base plates, 2.54 × 60.96 × 60.96 cm (1 × 24 × 24 in.), and the assembly was mounted on three legs made of 10.16 cm (4 in.) 2014-T6 I-beams. Portland cement concrete caps were cast around the top and bottom of each of the two columns.

Following the assembly of the columns, 16 thermocouples were attached to each column, three at each of four different levels as specified in accordance with ASTM E119; the top and bottom set were mounted 0.6 m (2 ft) from the ends, and the other two were equally spaced in between.

The two columns were then fireproofed by the common commercial practice of coating with lightweight vermiculite concrete, as shown in Fig. 2.1. The first column was wrapped with lath, and the vermiculite mix was applied to a depth of 5.08 cm (2 in.) from the face of the lath. The 5.08 cm measurement included a scratch coat 6.35 mm (¼ in.) thick, the base coat, and a finish coat of 1.5875 to 3.175 mm ($^1/_{16}$ to ⅛ in.) thickness of white lime plaster. The second column was wrapped with lath and the vermiculite mix was applied to a thickness of approximately 3.81 cm (1½ in.), including the scratch coat; it was then wrapped with a second layer of lath, and additional vermiculite was applied so that a total thickness of 7.62 cm (3 in.) was achieved, including the finish coat.

The fireproofing was done by McNulty Brothers of Chicago, a contractor experienced in the application of fireproofing materials to steel structures. With both columns, they used standard 1.5 kg (3.4 lb) diamond mesh lath, with keystone corner beads to ensure the indicated thicknesses of fireproofing were achieved.

The fireproofing materials were:

- Scratch coat: one 45 kg (100 lb) bag of glass-fibered gypsum to 0.05 m^3 (2 ft^3) of vermiculite aggregate
- Base coat: one 45 kg bag of glass-fibered gypsum to 0.07 m^3 (2½ ft^3) of vermiculite aggregate
- White finish coat: one 45 kg bag of unfibered gypsum to one 45 kg bag of lime

Bestwall glass-fibered gypsum (ASTM C-35) and Zonolite vermiculite (ASTM C-22) were used.

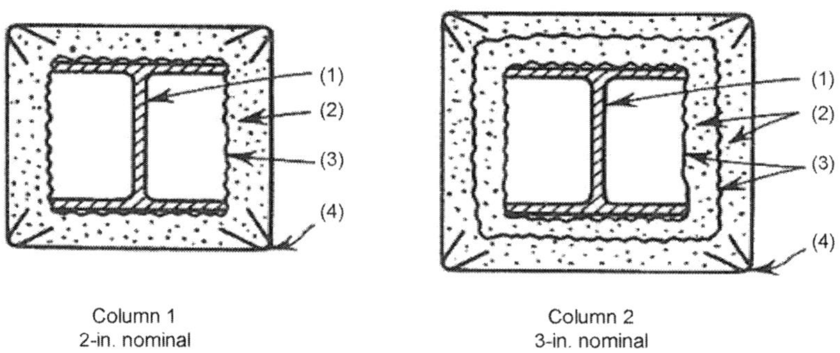

Fig. 2.1 Cross sections of aluminum columns ready for testing. (1) Aluminum alloy column. (2) Vermiculite concrete. (3) Lath layer. (4) Keystone corner beads. Source: Ref 2.2

After completion, the columns were allowed to harden and dry for four months before testing. The cross sections of the completed columns ready for testing are shown in Fig. 2.1.

The two fireproofed aluminum columns were tested individually by being placed in a 2.7 × 2.7 × 2.7 m (9 × 9 × 9 ft) gas-fired furnace and subjected to the standard ASTM time-temperature exposure cycle (Ref 2.3–2.5). The temperatures on each column were recorded throughout the tests on a Leeds & Northrup continuous potentiometer.

In the test of the column with 5.08 cm (2 in.) of vermiculite coating, the calcining period (the time for the water in the plaster to dry out) was approximately 65 min, after which the temperature of the column gradually increased at an increasing rate to a maximum of 13 °C (8 °F) per min. The average temperature of the column at the hottest level as measured by the thermocouples exceeded 190 °C (375 °F) in 2 h, 13 min and reached 260 °C (500 °F) in 2 h, 29 min (from the start of the test).

In the test of the column with 7.62 cm (3 in.) of vermiculite coating, the calcining period (the time for the water in the plaster to dry out) was approximately 2 h, and the maximum rate of increase of the temperature of the column was approximately 15 °C (5 °F). The average temperature of the column at the hottest level exceeded 190 °C (375 °F) in 4 h, 7 min and 260 °C (500 °F) after 4 h, 30 min (from the start of the test).

The average temperatures from the hottest regions in the columns are summarized in Fig. 2.2.

A cross plot of these data illustrating hours of protection as a function of the vermiculite coating thickness is shown in Fig. 2.3.

The data in Fig. 2.3 were then used to estimate the thicknesses of vermiculite needed for the protection of aluminum alloy structural members for various lengths of time as shown in Table 2.1. It is expected that simi-

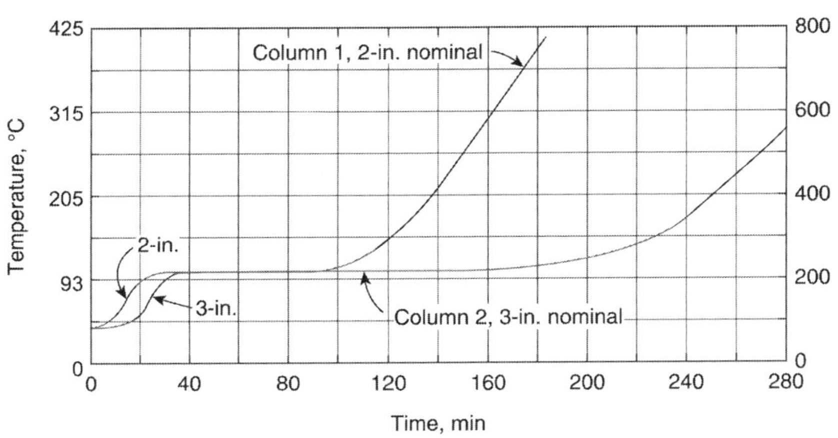

Fig. 2.2 Temperature in hottest region on aluminum beams during testing. Thickness of vermiculite coating on columns: Column 1, 5.08 cm (2 in.); column 2, 7.62 cm (3 in.). Source: Ref 2.2

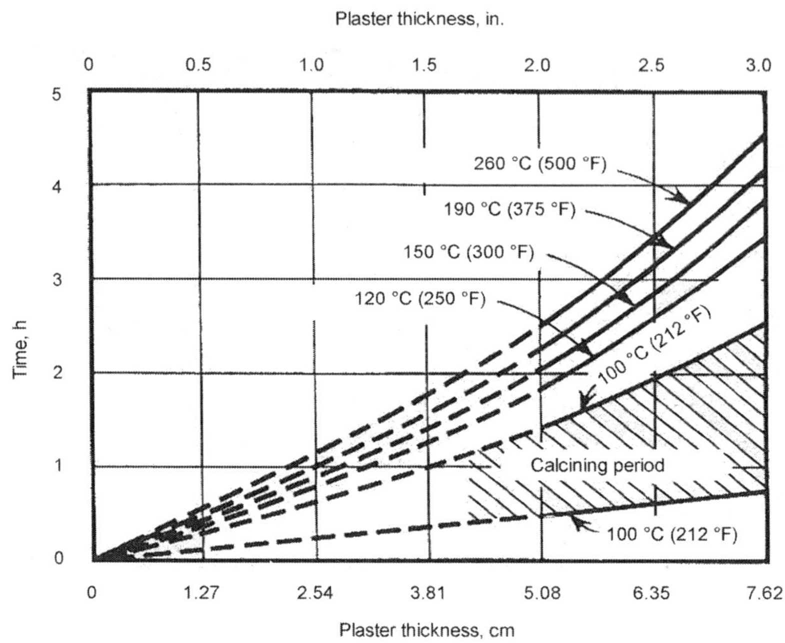

Fig. 2.3 Time to reach temperature with various thicknesses of vermiculite fireproofing. Source: Ref 2.2

Table 2.1 Calculated thickness of vermiculite (plaster) needed to fireproof aluminum members

Fire protection period, h	Plaster thickness(a) to prevent temperatures of Al columns in excess of:				Plaster thickness required for steel columns (540 °C, or 1000 °F)	
	150 °C (375 °F)(b)		260 °C (500 °F)(c)			
	cm	in.	cm	in.	cm	in.
1	3.175	1¼	2.858	1 1/8	1.905	¾
2	4.763	1 7/8	4.445	1¾	2.54	1
3	6.35	2½	6.033	2 3/8	3.493	1 3/8
4	7.62	3	7.303	2 7/8	4.445	1¾

(a) From the face of the lath. (b) To ensure no substantial change in properties at room temperature as a result of test exposure. (c) To ensure yield strengths at least equal to the design allowable stresses during the test exposure

lar thicknesses may also be used to provide fire protection for floors and roofs.

The data from these tests of aluminum columns can be reduced algebraically to permit calculation of fire protection periods for a wide range of column sizes and vermiculite coatings, to the following form:

$$R = c\left(\left(D^2 - ad^2\right)^n / D^n\right) \quad \text{(Eq 2.1)}$$

where

R = fire retardance period, h
D, d = outside and inside dimensions of the fireproofing, in.
n = constant equal to 1.7
c, a = constants dependent on the materials

The constants c and a may be calculated from the data for the aluminum columns as $c = 0.1$ and $a = 1.0$, so that the base equation reduces to:

$$R = 0.1((D^2 - d^2)^{1.7}/D^2) \qquad \text{(Eq 2.2)}$$

These test results indicate that the aluminum alloy structural members can be fireproofed by the same practical and commercially acceptable methods used for fire protection of steel structural members; the thicknesses of vermiculate coating are simply greater.

2.2 Rockwool Insulation for Fire Protection of Aluminum Naval Bulkheads

Lloyd's Register of Shipping and the American Bureau of Shipping have provided certification of their approval of mineral wool insulation of aluminum decks and bulkheads for use on ships classed and registered by those organizations (Ref 2.6, 2.7).

To determine the basis of their approval for this application, fire tests were conducted of a variety of ship structural components protected in various ways with mineral wool insulation. The tests were conducted by the Fire Insurers' Research & Testing Organization (FIRTO) in Borehamwood, Hertfordshire, England, and by the Norwegian Fire Research Laboratory in Avaldsnes, Norway, all in the 1983 to 1985 time frame (Ref 2.8, 2.9).

The aluminum bulkhead samples tested were (a) a 3025 × 3020 mm (10 × 10 ft) panel, reinforced on the exposed side by four continuous vertical angle stiffeners on 600 mm (2 ft) centers welded to the face plate of the bulkhead sample; (b) a 25,000 × 2500 mm (82 × 8.2 ft) bulkhead section with major joints; and (c) a deck floor section with underdeck insulation.

In all of these tests, the mineral wool insulation tested was Rockwool Firebatts 825 with a density of 110 ±10% kg/m^3 (6.867 ±10% lb/ft^3). In one set of tests, it was fitted to a minimum thickness of 80 mm (3 in.) on each side of the bulkhead plating to be protected; the 80 mm (3 in.) minimum was made up of two layers of at least 40 mm (1.6 in.) applied one layer at a time. Adjacent slabs of insulation were fitted tightly together, and the joints in the two layers were staggered to provide overlap. Where stiffeners protruded from the bulkhead plating, the extensions were covered with one 40 mm layer of the insulation.

Other mineral wool insulation combinations were also tested, some requiring two layers, one 40 mm (1.6 in.) plus another 25 mm (1 in.) in thickness.

In all cases, the layers of insulation were held in place with stainless steel pins. In the tests the minimum length of insulated bulkhead tested was 450 mm (18 in.).

Schematic drawings of typical insulated panels tested by the Fire Insurers' Research & Testing Organization (FIRTO) are shown in Fig. 2.4 and 2.5.

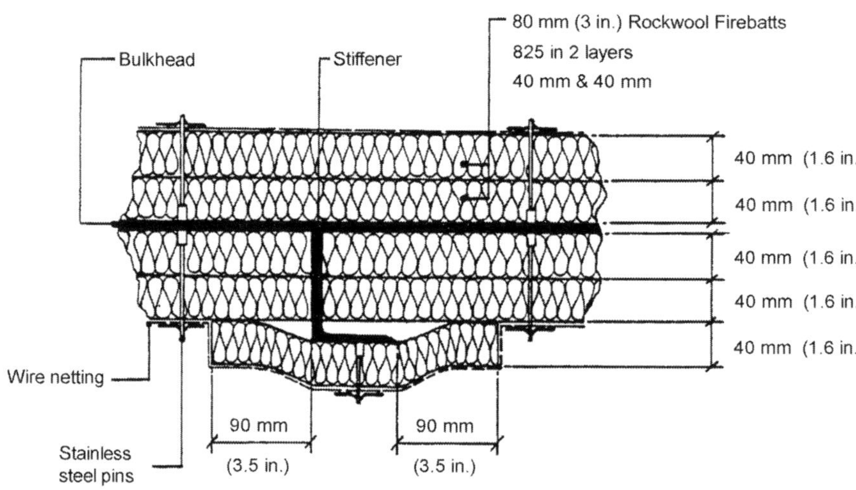

Fig. 2.4 Schematic drawing of typical section of mineral wool fire test sample with 80 mm (3 in.) thick protection. Source: Ref 2.6, 2.8

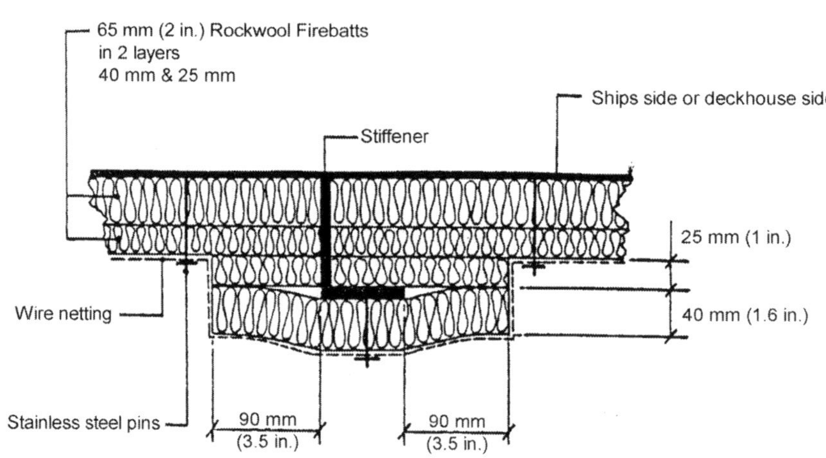

Fig. 2.5 Schematic drawing of typical section of mineral wool fire test sample with 65 mm (2 in.) thick protection. Source: Ref 2.6, 2.8

In the tests conducted by the Norwegian Fire Research Laboratory, the mineral wool insulation was in two layers, one 50 mm (2 in.) and the other 60 mm (2.3 in.; total 110 mm, or 4.3 in.). In these tests the much larger test sample was used: 25000 × 2500 mm (82 × 8.2 ft.). These tests included a bulkhead joint in the floor test of the configuration in Fig. 2.6.

All of these tests were conducted in accordance with British Standard 476, Part 7 (Ref 2.10) in oil-fired furnaces heated with a time-temperature curve intended to simulate a hydrocarbon fire. As a heating source, firing oil with a density of 0.83 kg/l (10.83 oz/gal) was used. The furnace tem-

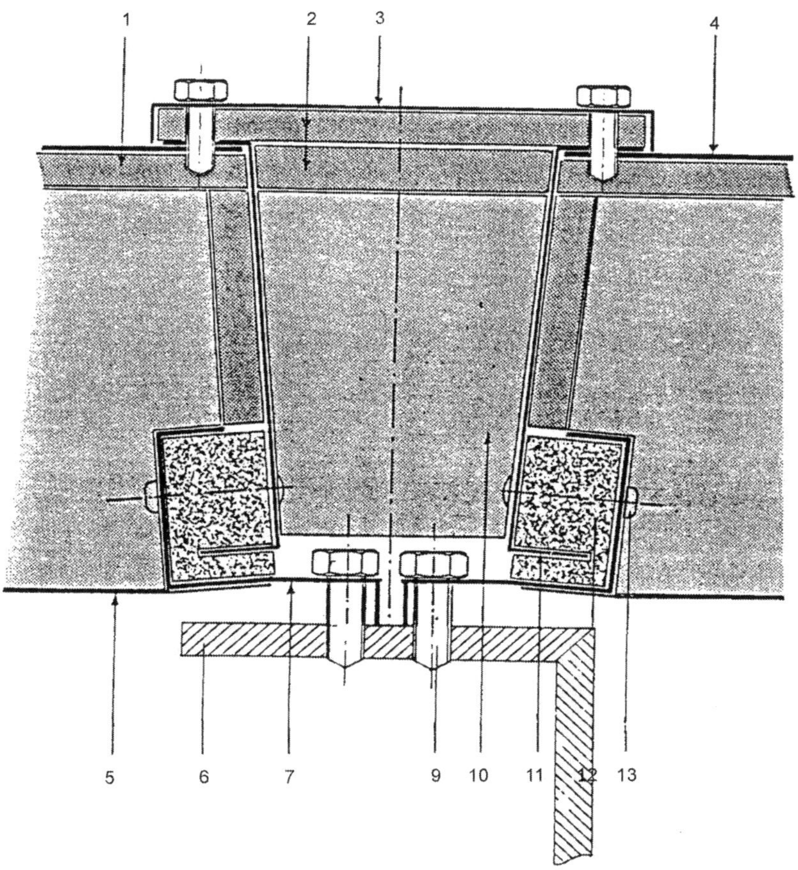

1–Rockwool insulation
2–Rockwool insulation
3–Stainless steel cap
4–Aluminum plate
5–Aluminum plate
6–Aluminum extruded shape
7–Stainless steel shim sheet
8–missing
9–Stainless steel bolt
10–Rockwool insulation
11–Stainless steel shim sheet
12–High-density Rockwool insulation
13–Pop rivet

Fig. 2.6 Schematic drawing of representative joint in floor and bulkhead test samples insulated with mineral wool. Source: Ref 2.6, 2.8

perature was recorded with the aid of four chromel alumel thermocouples protected by ceramic sleeves placed approximately 150 mm (6 in.) from the exposed surface of the test sample. The temperature of the test bulkhead was measured with copper/constant thermocouples each soldered to a copper disk and covered with asbestos pads.

The time-temperature cycle is:

Time, min	Temperature	
	°C	°F
0	20	72
3	880	1605
5	944	1731
10	1032	1890
15	1070	1960
30	1097	2005
60	1100	2012
120	1100	2012

The actual furnace temperature and the temperature of the aluminum core of the test bulkhead recorded as a function of time in a representative test are shown in Fig. 2.7.

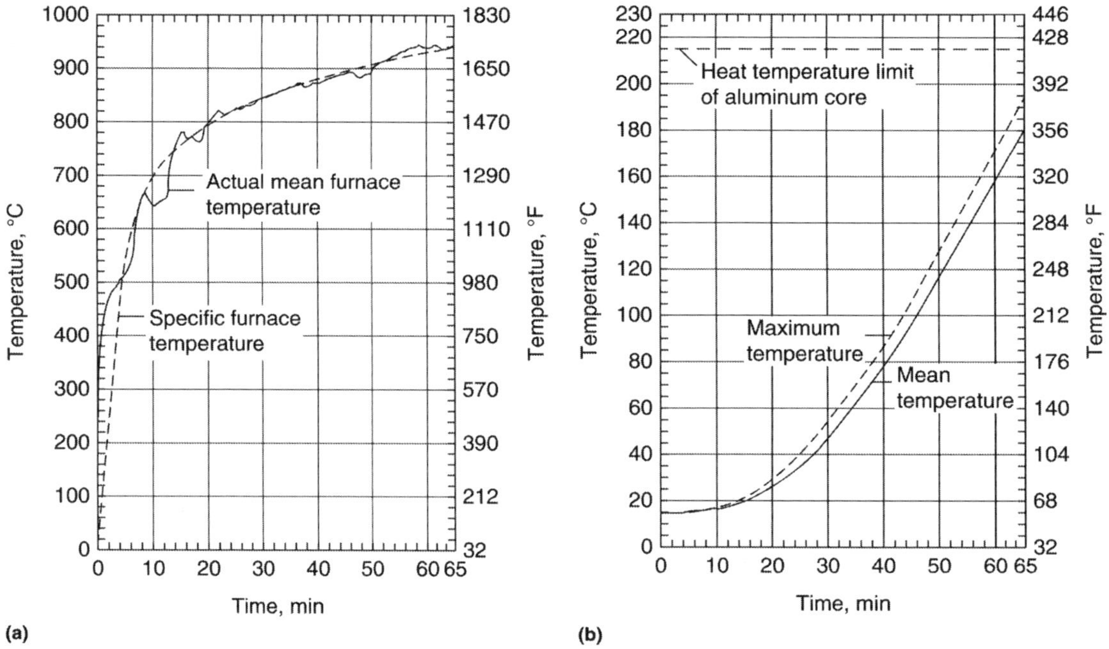

Fig. 2.7 Temperature records of (a) the furnace and (b) the aluminum core of a representative test sample during fire tests of mineral wool insulated aluminum bulkhead sections, FIRTO TE4741. Source: Ref 2.6

The information in the charts in Fig. 2.7 illustrates that (a) the furnace temperature followed the desired curve reasonably well, and (b) the mean and maximum temperatures of the aluminum core rose 146 and 159 °C (295 and 318 °F), respectively, not exceeding the specified limit of 200 °C, and therefore conforming to the A60 fire test requirement.

Other observations made during the tests were:

- Ten minutes after the test started, the exposed face of the specimen appeared incandescent.
- After 60 min, the aluminum bulkhead showed no visible distortion.
- After 65 min, the test was ended, with no damage to or distortion of the test sample.

The overall conclusion from all tests of aluminum bulkhead and deck sections was that the bulkhead protected as in the test "prevented the passage of smoke and flame and satisfied the A60 classification" of fire protection for ship bulkheads. As a result of these test results, the American Bureau of Shipping and Lloyd's Register of Shipping has accepted this mode of insulation for aluminum bulkheads in ships (Ref 2.6, 2.7) since 1985.

2.3 General Guidelines for Fire Protection of Ship Structures

To provide guidelines comparable to those previously established for steel, the Society of Naval Architects and Marine Engineers published a Technical and Research Bulletin 221 in 1974 entitled "Aluminum Fire Protection Guidelines" (Ref 2.9). This Bulletin describes two methods for designing the fire protection of aluminum structures in ships: the traditional Class System used for steel, and an approach that bases fire protection on the fire exposure.

The Bulletin is based on a series of fire tests conducted at the National Bureau of Standards in accordance with ASTM Standard E119 (Ref 2.3–2.5). The tests established the amount of insulation needed to protect aluminum structures from loss of structural integrity for various lengths of time under a standard fire exposure.

In 1976, the National Bureau of Standards published a report entitled "Fire Performance Testing of Bulkhead Insulation Systems for High Strength-to-Weight Ship Structures" (Ref 2.11) for the Ship Damage Prevention and Control Section of the Naval Sea Systems Command. This document reports on 16 insulated aluminum bulkhead specimens, two insulated and two unprotected glass reinforced plastic specimens, and intumescent painted aluminum and steel panel specimens. The tests were conducted with a 0.6 m (2 ft) horizontal slab furnace, which is suitable for screening out less promising candidate materials with regard to their fire endurance. The specimens insulated with either refractory fibrous material or with mineral wool gave the best overall performance.

2.4 Other Options for Fire Protection of Aluminum Structures

Other options available for the fire protection of aluminum alloy structural members are:

- Ceramic fiber of high-purity aluminum and silica may be pressed into boards or mats for encasing structural members. Examples of these products are Fiberfrax, Ceramaterials, and Gemcolite (Ref 2.12).
- Calcium silicate boards consisting of calcium oxide and silica combined with vermiculite are used as laminates to protect structural members. Examples of these products are Vermiculux from Invicta, Super Firetemp from Red Seal Co., and Skamotec from Skamol (Ref 2.13).
- Gypsum boards, consisting of calcium sulfates bonded with water, may be used to encase structural members. Insulation is provided by the time required to release and evaporate the water. Guidance in the use of gypsum is available from the Gypsum Association (Ref 2.14).
- Intumescent materials—those that undergo a chemical change when exposed to heat, becoming viscous, and expanding to form bubbles that harden into a multicellular structure—are also used in sheathing to provide fire protection. They usually contain graphite compounds and may be applied as paints, putty, foams, or strips. Examples of these products are FireGuard E-84 paint, Astroflame, coatings (Ref 2.15).

REFERENCES

2.1 S. Lundberg, "Material Aspects of Fire Design," TALAT Lecture 2502, European Aluminium Association, 1994

2.2 J.G. Kaufman and R.C. Kasser, Fire Protection for Aluminum Alloy Structural Shapes, *Civil Engineering,* March 1963, p 46–47

2.3 "Standard Methods of Fire Tests of Roof Coverings," ASTM E108, *Annual Book of ASTM Standards,* ASTM (updated annually)

2.4 "Standard Test Methods for Fire Tests of Building Construction and Materials," ASTM E119, Part 04.07, *Annual Book of ASTM Standards,* ASTM (updated annually)

2.5 "Standard Test Method for Behavior of Materials in a Vertical Tube Furnace at 750 °C," ASTM E136, *Annual Book of ASTM Standards,* ASTM (updated annually)

2.6 "Aluminum Bulkhead—Insulated With 80 mm Rockwool Firebatts 825 on Both Faces," Certificate No. ICD/F83/697, Lloyd's Register of Shipping, London, Dec 21, 1983; also Reference MHC/MT, March 29, 1985

2.7 "Rules for Building and Classing Aluminum Vessels," Notice No. 5, American Bureau of Shipping, 1975

2.8 SINTEF Test Certificate 250000.20/86.020, Hydro Aluminum Offshore A/S, Norwegian Fire Research Laboratory, Dec 17, 1986

2.9 "Aluminum Fire Protection Guidelines," Technical and Research Bulletin 2-21, Society of Naval Architects and Marine Engineers (SNAME), Jersey City, NJ, 1974

2.10 "Classification of Materials for Fire Resistance, Part 7," British Standard 476, The British Standards Institution

2.11 "Fire Performance Testing of Bulkhead Insulation Systems for High Strength-to-Weight Ship Structures," National Bureau of Standards, 1976

2.12 Thermal Products Co., www.products.thermalproducts.com, Cera-Materials Co., www.ceramaterials.com/ceramicfiberboards.html; Refractory Specialties, Inc., www.rsifibre.com/products/

2.13 Invicta Fire Protection, www.durasteel.net; Promat Co., www.promat.co.uk/en/products/vermiculux; Red Seal Electric Co., www.redeal.com/non-metallic-insulation-products/firetemps-sfl.wspx; Skamol, www.skamotec225.us

2.14 "Fire Resistance Provided by Gypsum Board Membrane Protection," GA-610-13, Gypsum Association, Hyattsville, MD, 2013

2.15 Shield Industries Co., www.shieldindustries.com/fireguard_wp; Astroflame (Fire Seals) Ltd., www.astroflame.com/intumescent-paint/index.html

CHAPTER 3

Aluminum in Fire-Sensitive Applications

ALUMINUM has been thoroughly evaluated and successfully used for a great variety of components and structures where the concern of fire hazards and/or the need for fire protection may be encountered. The sections that follow describe several of the more important applications of aluminum alloys and products where fire resistance is useful. The chapter concludes with two historical accounts of aluminum behavior in fires. It is interesting to note that in these incidents, although aluminum was originally thought to contribute to the spread of the fires, there was actually no evidence to substantiate the accounts.

3.1 Offshore Oil Rigs

In connection with consideration of its use for both oil rigs and building structures, aluminum was tested in accordance with British Standard 476, Parts 3 and 4 (Ref 3.1, 3.2), and, as noted previously, received the highest ratings in each case. Specific design rules and guidelines for the use of aluminum in the superstructures of such fire-critical applications as oil rigs and their incorporated helidecks are covered in Ref 3.3 to 3.10. (See Fig. 3.1.)

Neelish Uppal, Aluminium Structures Pte Ltd, published an overview of the advantages of using aluminum in offshore oil rigs and other structures (Ref 3.11). Among the broad range of subjects covered, those pertinent to the subject of this book include the thermite sparking and burning of aluminum. Uppal indicates that perhaps the greatest fear of offshore

Fig. 3.1 Offshore oil rig with aluminum superstructure. Source: Ref 3.4–3.6, 3.11

engineers in using aluminum components is the possibility of an explosion resulting from an exothermic sparking reaction between rusty steel and aluminum, and the necessity of painting any aluminum surface that might experience contact with rusty steel.

In Ref 3.11, Uppal itemizes the reasons why the use of aluminum alloys should not cause concern, as explained in Chapter 1, "Properties and Characteristics of Aluminum and Aluminum Alloys," in this book. Aluminum alloys have:

- High thermal conductivity, approximately four times that of steel, that moves the concentrated heat quickly away
- Higher specific heat, twice that of steel
- High reflectivity, many times higher than that of steel
- The modulus of elasticity, approximately one-third that of steel, requiring designs to be based on deflection, therefore providing a considerable strength safety factor for aluminum compared to a typical steel design

Bayley (Ref 3.3) also reports the advantages of the use of aluminum alloys for the topside structures of oil rigs, and notes an additional option for the use of aluminum sheeting as a sacrificial external coating for load-bearing columns in the case of fire. As temperatures increase in a nearby fire, the outer aluminum skin melts, revealing a supported fire insulation material that provides the required period of fire performance and protects the structural aluminum elements. See Chapter 2, "Fire Protection of Aluminum Structures," in this book.

3.2 Building Structures

For building walls and skin structures, covered by BS 476, Part 3 (Ref 3.1), aluminum has been found to have both the highest resistance to fire penetration and flame spread and was classified as AA. See Fig. 3.2. For inner surfaces, covered by BS 476, Part 4 (Ref 3.2), aluminum demonstrated a very high level of resistance to flame spread and was classified as 0, the highest rating for that type of assembly. Parts 5 and 6 of BS 476 (Ref 3.12 and 3.13, respectively) provide additional ratings for materials of construction. Part 5 provides for a test for ignitability: materials are rated "easily ignitable," X, or "not easily ignitable," P. Aluminum received a P rating. Part 6 of BS 476 covers fire propagation performance for coated systems; because of its high corrosion resistance and the need for only thin coatings, aluminum consistently achieved high ratings in this situation as well.

The results of European Community consideration of the selection of materials of construction are given in Ref 3.14. Aluminum and aluminum alloys are classified as Class A, "without the need for testing" except when it is used in a finely divided form (e.g., powder).

In the United States, aluminum clad building materials such as REYNOBOND (Alcoa Architectural Products) have been evaluated for a number of years in flame spread and fire resistance tests in accordance with ASTM Standards E108 (Ref 3.15) and E119 (Ref 3.16), and UBC 17-5 and 17-6,

Fig. 3.2 Examples of the use of aluminum for internal and external building construction. Source: Ref 3.10

their full-scale multi-story fire test, receiving fire-resistant ratings (Ref 3.17).

Alucobond, a product of 3A Composites GmbH, of Singen, Germany, has received worldwide approval of its aluminum-mineral-filled core panel, as noncombustible based on the provisions of British Standard 476 and many other international rating standards agencies (Ref 3.18).

3.3 Over-the-Road Vehicles

Aluminum has been successfully used in tank trucks to haul gasoline, diesel fuel, and other flammable and combustible liquids since the 1950s. See Fig. 3.3. The light weight of aluminum combined with its flexibility in design, fabricating, and maintenance have made it the material of choice for such applications. The nonsparking characteristic of aluminum is also an attractive feature for flammable liquid tanks. From the regulatory standpoint, aluminum has been recognized by the federal government for approximately 40 years; National Highway Transportation Safety Board (NHTSA) rules MC304 and MC305 have referenced aluminum since at least December 31, 1955 (Ref 3.19).

Many thousands of aluminum gasoline tank trucks are in service, and with such exposure there are inevitable accidents, some resulting in fires. The temperature of a gasoline vapor fire (approximately 925 °C, or 1700 °F) invariably causes the aluminum to melt down as the liquid boils off. This is a safety feature, because explosions that might injure firefighters are not possible. Such fires are dramatic, however, and frequently are investigated by the appropriate authorities. No incident has been reported in which it has been alleged that the aluminum "burned" to contribute to the severity of the fire.

The high toughness of the aluminum tank structures, usually of aluminum alloy 5454, also assists in containing the fluids because it resists penetrating damage and minimizes any health exposure.

Aluminum alloys are also widely used in automotive and truck applications. First widely introduced in Audi automobiles, aluminum alloys are

Fig. 3.3 Oil tank truck with aluminum body. Source: Ref 3.10, 3.17

increasingly found in most autos and light trucks for the advantage they provide in mileage (Fig. 3.4–3.7).

3.4 Railroad Cars

Aluminum is also used for a number of railroad car platforms and bodies, including those that operate carrying hot contents long distances. For example, aluminum alloys of the 5*xxx* (aluminum-magnesium) series are commonly used for molten asphalt tanks for temperatures up to 190 °C (375 °F), as shown in Fig. 3.8.

High-strength structural alloys are used in the undercarriage of passenger rail cars, while bright finish alloys provide the exterior (Fig. 3.9).

Fig. 3.4 Aluminum alloys make commercial vehicles such as buses more fuel efficient. Source: Ref 3.10, 3.17

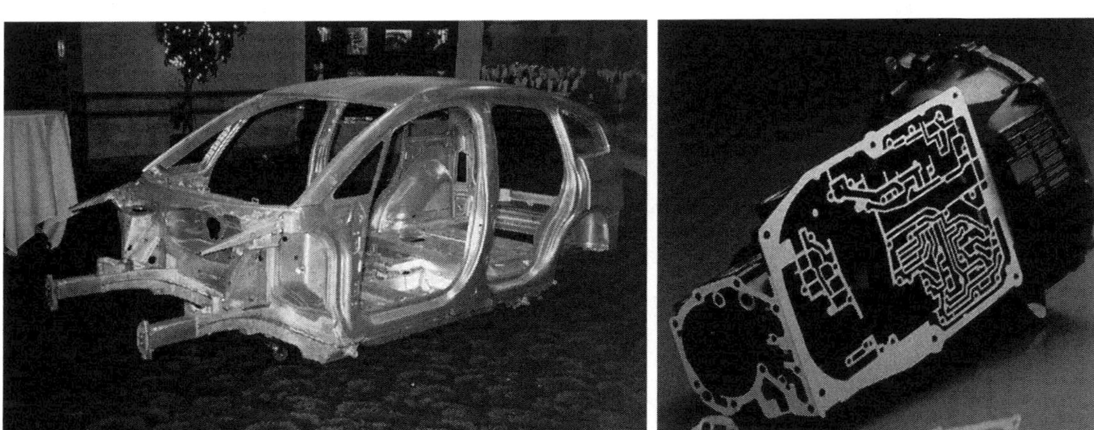

(a) (b)

Fig. 3.5 (a) Aluminum is safely used in entire car body frames. (b) Complex aluminum alloy castings find their way into many engine components where they safely help conduct heat away. Source: Ref 3.18

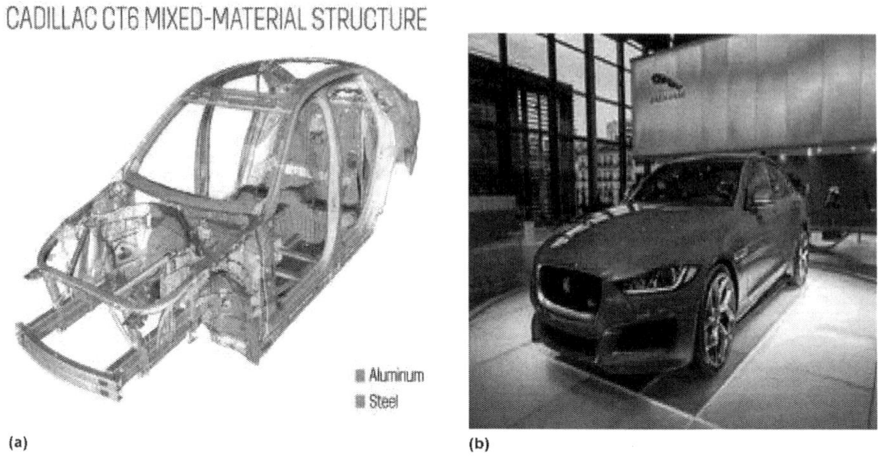

Fig. 3.6 (a) Cadillac CT6 and (b) Jaguar XE make good use of aluminum alloys in their body structures.

Fig. 3.7 The Ford F-150 has lead the way for heavy-duty trucks in light-weighting with aluminum alloys.

Fig. 3.8 All-aluminum hopper cars safely carry coal and hot asphalt. Source: Ref 3.10

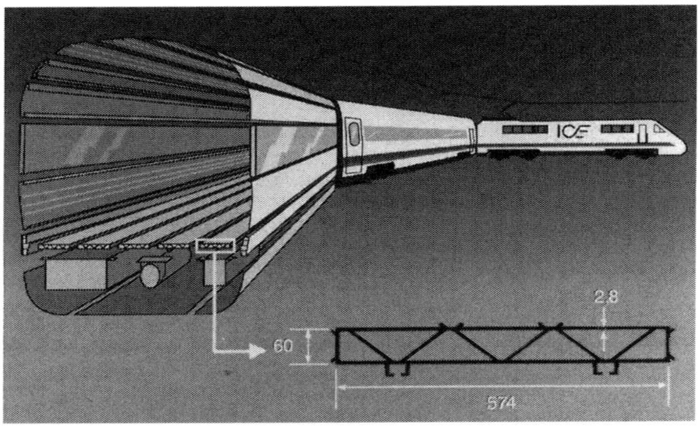

Fig. 3.9 Passenger car framing and undercarriage make good use of aluminum extruded shapes. Source: Ref 3.10

3.5 Commercial Ships

The most significant applications of aluminum alloys in commercial marine construction have been in (a) the superstructures of cruise ships (Fig. 3.10), (b) the construction of relatively small ships for personal or commercial use, and (c) more recently, the complete hull and superstructure of fast ferries (Fig. 3.11). The large number of aluminum vessels of various sizes that have gone into service over the past 75 years reflects the confidence that aluminum alloys have built up in the commercial marine field (Ref 3.3–3.10, 3.20, 3.21).

This confidence was confirmed by the American Bureau of Shipping (ABS) in its documents covering aluminum alloys, the first "Rules for Building and Classing Aluminum Vessels," issued in 1975 (Ref 3.20), and the current updated version published in 2014 (Ref 3.22). In these documents, the ABS provides a list of approved alloys with their design properties. The wrought alloys endorsed in the ABS publications include: strain-hardenable alloys 5052, 5059, 5083, 5086, 5383, 5454, 5456, and 5754; and heat treatable alloys 6005, 6005A, 6061, 6063, and 6082. All of these alloys are approved for use when welded; weld wire alloys 4043, 5183, 5356, 5554, and 5556 are approved. Castings of alloys 356.0, A356.0, and 357.0 are also approved, including when welded with 4043 and 5356.

The International Convention for the Safety of Life at Sea (SOLAS) (Ref 3.23) classifies aluminum and its alloys as noncombustible materials. It expressly permits the use of aluminum alloys in ship construction, noting that a "suitably insulated aluminium alloy" would be expected to perform the same as steel. Rules from SOLAS for the protection of aluminum alloys are essentially the same as for steels: Type A ship divisions (a) must be constructed to prevent the passage of smoke and flames until the end of

Fig. 3.10 Cruise ship superstructures make good use of aluminum sheet, plate, and extruded shapes. Source: Ref 3.18

Fig. 3.11 All-aluminum fast ferries move passengers quickly and safely. Source: Ref 3.10

a one-hour test according to Resolution A.754, and (b) must be insulated with approved noncombustible materials.

All-aluminum commercial vessels, from oceangoing trailer ships to smaller fishing boats, have been in successful use for many years. More recently, the broad use of aluminum in high-speed "fast ferries" has greatly expanded around the world, notably in Australia and the Asian Rim. These vessels must and do meet all existing codes for fire resistance and for protection of passengers in fire emergencies (Ref 3.3–3.7, 3.20, 3.22–3.24).

The use of aluminum in the entire structure of fast ferries has become the standard (Ref 3.25). An illustration of its application for the internal structure of a fast ferry is shown in Fig. 3.12.

(a) (b)

Fig. 3.12 (a) Internal stiffening structure and (b) hull framing of an all-aluminum fast ferry. Source: Ref 3.10

3.6 Naval Vessels

The requirements imposed on the design of U.S. naval vessels by sophisticated radar, gun control, and communications systems have increased the size and weight of the superstructures. In order to maintain the required level of transverse stability, either topside weight must be reduced or the beam of the ship increased. Aluminum has been the traditional answer for most major surface combatants because of its advantageous strength-to-weight ratio.

3.6.1 Specific Applications

Aluminum alloys have been widely used in the superstructures of cruisers, destroyers, and frigates (Ref 3.26). They are also used in the hulls as well as superstructures of such high-performance craft as gun boats, patrol boats, surface effect ships, air cushion landing craft, and hydrofoils. In addition to use in the hulls and superstructures of these vessels, aluminum alloys are also used internally in all types of vessels for duct work, nonstructural partitions, doors, masts, and electronic equipment housings (Ref 3.7, 3.20, 3.22).

Alloys 5086 and 5456 in the H116 condition are the structural aluminum materials predominately used by the U.S. Navy. The H116 temper of these alloys is a special temper developed by the aluminum industry to provide a material having improved resistance to exfoliation and stress corrosion (Ref 3.27). Alloy 5454 is used in areas adjacent to stacks because of its favorable elevated-temperature mechanical properties.

Aluminum alloy designs are in the competition for faster Navy vessels with exceptional speed and range, such as the Austal/General Dynamics trimaran in Fig. 3.13.

The choice of aluminum still raises controversy from some quarters because of lack of understanding of how aluminum behaves in an engulfing fire (it melts and runs off) and the necessity of buffering connections between aluminum and steel members (direct contact leads to galvanic corrosion). But the lightweighting provided by aluminum either in the superstructure alone or in the entire ship (as in fast ferries) enables faster speeds, longer ranges, and in some designs, greater load-carrying capacity.

3.6.2 Fire Protection of Aluminum Naval Vessels

According to representatives of the U.S. Navy, a program for fire protection of aluminum structures began after the collision of the USS *Belknap* and *Kennedy* in 1975 (Ref 3.28, 3.29). It is understood that when the vessels collided, jet fuel lines adjacent to the *Kennedy* flight deck were ruptured and aviation fuel poured onto the *Belknap.* It was ignited purportedly by arcing of electrical wiring that was damaged by the collision. Reportedly, the entire superstructure of the *Belknap* was engulfed in flames continually fed by the ruptured fuel lines on the *Kennedy,* with resulting extensive damage.

Fig. 3.13 Austal/General Dynamics trimaran design for the U.S. Navy

The purpose of the fire protection program was to determine measures needed to protect aluminum for a period of approximately 30 min, which the Navy considers long enough for fire control measures to be initiated.

In order to qualify fire protection systems that would provide 30 min of protection, the Navy used the ASTM E119 test method (Ref 3.16). This is a test method that has been used for over 60 years for determining the fire-resistance ratings of assemblies of materials used in building construction. The fire exposure is considered to be representative of the way fires progress in confined spaces.

Fire tests for qualifying fire protection systems were conducted at the National Gypsum Co. (Ref 3.28) facilities in Buffalo, N.Y., and at the Philadelphia Naval Shipyard (Ref 3.29). The tests at National Gypsum were based on the ASTM E119 procedures and were used for screening the systems. The tests performed at the Philadelphia Naval Shipyard were tests of full-scale sections fabricated from portions of the *Belknap* superstructure.

Refractory fiber thermal insulation material such as Fiberfrax (Unifrax I LLC) was qualified for use under these tests and military specifications covering passive fire insulation materials. Aluminum alloys performed as expected in all tests with such materials, achieving satisfactory residual strength and structural integrity. As a result, the Navy program established procedures for fire protection of aluminum structures on existing vessels as well as new ones. Retrofitting thermal insulation on existing naval craft involves gutting the superstructure, and the final product is considered comparable to installation of such insulation on a new vessel.

3.6.3 Accounts of Aluminum Behavior in Fires

As noted in Chapter 2 in this book, "Fire Protection of Aluminum Structures," much of the concern about the possibility that aluminum may burn and contribute to flame spread in a serious fire arose from inaccurate reports about the 1982 sinking of four British Royal Navy warships off the Falkland Islands in the South Atlantic during what is generally referred to as the Falklands War (Ref 3.30), and again later with the 1987 damage to the U.S. Navy *Stark* guided-missile frigate in the Persian Gulf (Ref 3.31, 3.32). These incidents are discussed in more detail subsequently.

The Falklands War. Worldwide major print and electronic media coverage of the 1982 sinking of four British Royal Navy warships reported that the use of aluminum in the ships' structures and the burning of that aluminum during the fires generated by the bombing contributed to and hurried the loss of the ships. This occurred despite the fact that the ships that sunk were made entirely of steel, and their loss had nothing to do with aluminum.

One commentator attributed the loss of the HMS *Sheffield* by fire to the widespread use of aluminum, which he said "burned easily." That report was picked up by many press services and repeated around the world, leading to concerns about the use of aluminum in any applications potentially subject to fire. Three more ships were sunk following the *Sheffield,* including the HMS *Ardent,* HMS *Antelope,* and HMS *Coventry,* and many falsely jumped to the conclusion that all of these succumbed because of aluminum burning.

Following the sinking of the ships and with concerns about their veracity being raised, the British Admiralty and a number of other organizations conducted detailed studies of the events. Following the British Admiralty study, the Secretary of State for Defense of the United Kingdom presented a detailed analysis of all aspects of their conduct of the Falklands War in a report "The Falklands Campaign: The Lessons," dated December 1982 (Ref 3.30). In the discussion related to warships, the report noted that:

- There was no aluminum used in the HMS *Sheffield,* which was the focus of most of the inaccurate reports.
- The use of aluminum in naval ships was limited because of its low melting point.
- There were "significant savings" in the use of aluminum in warships above the waterline because of its light weight.
- The conclusion of the British Admiralty was "there is no evidence that it (aluminum) has contributed to the loss of any vessel."

Military analysts emphasized that the sophistication and destructive powers of the missiles and other weapons, along with the lack of adequate defensive measures, caused the loss of the vessels.

Damage to The USS *Stark*. Late in the day on May 17, 1987, an Iraqi F-1 Mirage aircraft fired two Exocet missiles at the U.S. Navy guided missile frigate USS *Stark* (FFG 31). Approaching the USS *Stark* at near the speed of sound, the first missile punched through the hull near the port bridge wing, burning at 1925 °C (3500 °F), and embedded in the interior but failed to explode. The second missile exploded on contact and added to the scope of the fire.

Investigation of the USS *Stark* incident showed that there was no evidence that aluminum had burned or contributed in any way to the damage, the cause of which was actually the highly incendiary makeup of the Exocet missile itself. A navy spokesman noted that at the temperatures generated in the USS *Stark* by the missile "steel and aluminum (behave) pretty much the same" (Ref 3.31, 3.32).

REFERENCES

3.1 "Classification of Materials for Fire Resistance, Part 3: External Fire Exposure Roof Test," British Standard 476, The British Standards Institution, 1975

3.2 "Classification of Materials for Fire Resistance, Part 4: Non-combustibility Test for Materials," British Standard 476, The British Standards Institution, 1970

3.3 M.J. Bayley, Application of Aluminum to Offshore Topside Structures, *Proceedings of the First International Offshore & Polar Engineering Conference,* Edinburgh, United Kingdom, Aug 11–16, 1991

3.4 "Basic Guide on the Use of Aluminium in Offshore Structures," Aluminum Design Guide, Vol 1, Section 1.4, Fire and Safety, Alcan Offshore, Wimpey Offshore, Nov 1990

3.5 "Basic Guide on the Use of Aluminium in Offshore Structures," Aluminum Design Guide, Vol 2, Section 4.4, Alcan Offshore, Wimpey Offshore, Nov 1990

3.6 "Fire Performance of Aluminium" (based on British Standard BS 476), Alcan Offshore, Wimpey Offshore, Nov 1990

3.7 "Fire Protection," Offshore Standard DNV-OS-D301, Det Norske Veritas, 2013

3.8 D.G. Altenpohl, Ed., *Aluminum: Technology, Applications and Environment,* The Aluminum Association, Washington, D.C., and TMS, Warrendale, PA, 1997

3.9 J. Gilbert Kaufman, *Introduction to Aluminum Alloys and Tempers,* ASM International, 2000

3.10 J. Gilbert Kaufman and E.L. Rooy, *Aluminum Alloys Castings: Properties, Processes, and Applications,* ASM International, 2004

3.11 N. Uppal, The Structural Use of Aluminium with Particular Reference to the Offshore Industry, *Proceedings of Alumitech'97,* May 1997

3.12 "Classification of Materials for Fire Resistance, Part 5: Ignitability of Building Materials" (now replaced by Part 4), British Standard 476, The British Standards Institution

3.13 "Classification of Materials for Fire Resistance, Part 6: Fire Combustibility of Coated Systems" (now obsolete), British Standard 476, The British Standards Institution

3.14 "94/611/EC: Commission Decision of 9 September 1994 Implementing Article 20 of Directive 89/106/EEC on Construction Products," European Economic Community, Sept 9, 1994

3.15 "Standard Methods of Fire Tests of Roof Coverings," ASTM E108, *Annual Book of ASTM Standards,* ASTM (updated annually)

3.16 "Standard Test Methods for Fire Tests of Building Construction and Materials," ASTM E119, Part 04.07, *Annual Book of ASTM Standards,* ASTM (updated annually)

3.17 Aluminum Composite Building Panel from Reynolds, *Metal Bulletin Monthly,* n222, June 1989

3.18 "Alucobond A2 Non-Combustible," 3A Composites GmbH, Singen, Germany, circa 2000

3.19 Rules MC-304+305, National Highway Transportation Safety Board (NHTSA), Washington, D.C. Dec. 31, 1955
3.20 "Rules for Building and Classing Aluminum Vessels," Notice No. 5, American Bureau of Shipping, 1975
3.21 Aluminium, the Marine Metal, *Alcan Marine,* Chapter 1, p 9–20
3.22 "Rules for Materials and Welding—Part 2: Aluminum and Fiber Reinforced Plastics (FRP)," ABS 0022:2014, American Bureau of Shipping, Houston, TX, 2014, Chapters 5–6
3.23 "International Convention for the Safety of Life at Sea (SOLAS)," International Maritime Organization, 1974
3.24 "Fire Performance Testing of Bulkhead Insulation Systems for High Strength-to-Weight Ship Structures," National Bureau of Standards, 1976
3.25 "Application of Aluminum Alloys to Fast Ferries," Publication No. 77, Aluminum Association, Washington, D.C., 1997
3.26 R.A. Sielski, The History of Aluminum as a Deckhouse Material, *Nav. Eng. J.,* May 1987, p 165–172
3.27 C.L. Brooks, Aluminum Magnesium Alloys 5086 and 5456 H116, *Nav. Eng. J.,* Aug 1970
3.28 "Aluminum Fire Protection: Results of Small Scale Panel Fire Test," Gibbs and Cox Inc., Sept 30, 1976
3.29 "Fire Endurance Tests for Naval Ship Engineering Center," Naval Ship Engineering Center (NSEC), Philadelphia, May 16–June 2, 1977
3.30 "The Falklands Campaign: The Lessons," presented to Parliament by the Secretary of Defence by Command of Her Majesty, Dec 1982
3.31 "Formal Investigation into the Circumstances Surrounding the Attack on the USS *Stark* (FFG 31) on 17 May 1987," Report 5102 Ser00/s-0487, by the Commander, Cruiser-Destroyer Group Two, FPO Miami, FL, June 12, 1987
3.32 R. Peniston, USS *Stark,* On Fire, *No Higher Honor,* www.navybook.com/no-higher-honor/timeline/uss-stark-on-fire/ (accessed Jan 3, 2016)

CHAPTER **4**

Estimating the Properties of Aluminum Alloys Exposed to Fire

ONE OF THE MOST IMPORTANT TASKS after a fire damages a building or structure of any kind but leaves it standing in place is to estimate how much the load-carrying capacity has been compromised by the heat of the fire. Questions to ask include: Will the building be able to stand future loadings, from both static and dynamic forces? How much has the load-carrying capacity of the structure been reduced?

There are two methods available to help analyze aluminum alloys in these cases: hardness tests and electrical conductivity measurements. Neither of these processes provide high-precision measurements of tensile or yield strength, but through well-established correlations, determinations can be made of whether or not damage has occurred and, if so, estimate how much the strength has been compromised.

4.1 Hardness Tests

Hardness testing offers significant means of estimating the strength of an aluminum member under any condition, including after it has been exposed to heat from a fire or other source. Of considerable help is the fact that various types of portable hardness machines are available for use when the measurements must be made in situ to avoid removing test samples from a structure that is potentially still active.

Some of the types of portable hardness testers that are useful for testing components in situ following a fire are described subsequently.

Webster hardness testing pliers combine reproducible test results with ease of operation (Fig. 4.1).

The standard Webster hardness tester enables measurements to be made on material thicknesses up to 6 mm (¼ in.); a modified tester enables testing of materials up to 25 mm (1 in.) thickness. It has the advantage of having been developed specifically for the purpose of estimating the effects of fire damage. An additional advantage is that its operation is essentially independent of the skill of the operator.

The Webster Instrument Company provides the conversion chart in Fig. 4.2, enabling easy conversion of Webster values to the Rockwell E scale, which, in turn, can be related to tensile strength by the type of conversion illustrated in the subsequent section "Relationship to Tensile Strength."

Other portable hardness testers Include the following three types:

- *The Barcol impressor,* a handheld portable hardness tester (Fig. 4.3), is especially suitable for testing flat and lightly curved surfaces. Conversion charts are also available, providing conversion of Barcol hardness test values to Rockwell, Vickers, and Brinell hardness values (Fig. 4.4).
- *The portable Sclerograph hardness tester* (Fig. 4.5) uses the rebound method and is suitable for determining the hardness of steel, nonferrous metals, and rubber.
- *The portable PZ3 Brinell testing clamp* (Fig. 4.6) hardness tester is suitable for standardized static Brinell ball-indentation tests up to 3000 kg. The PZ3 Brinell clamp has the advantage that the values determined are considered to be directly relatable to Brinell hardness values from full-size machines.

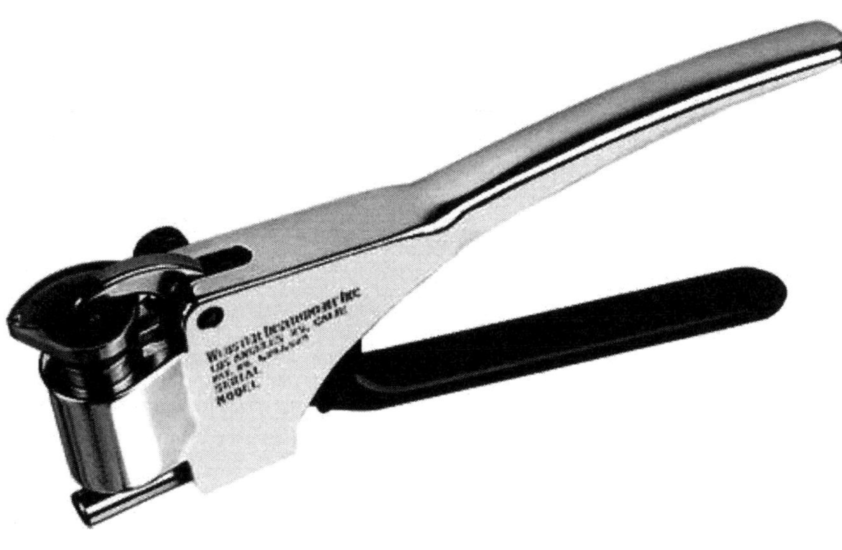

Fig. 4.1 Webster Hardness Tester Model B for aluminum and aluminum alloys. Printed with permission of Webster Instruments, Inc.

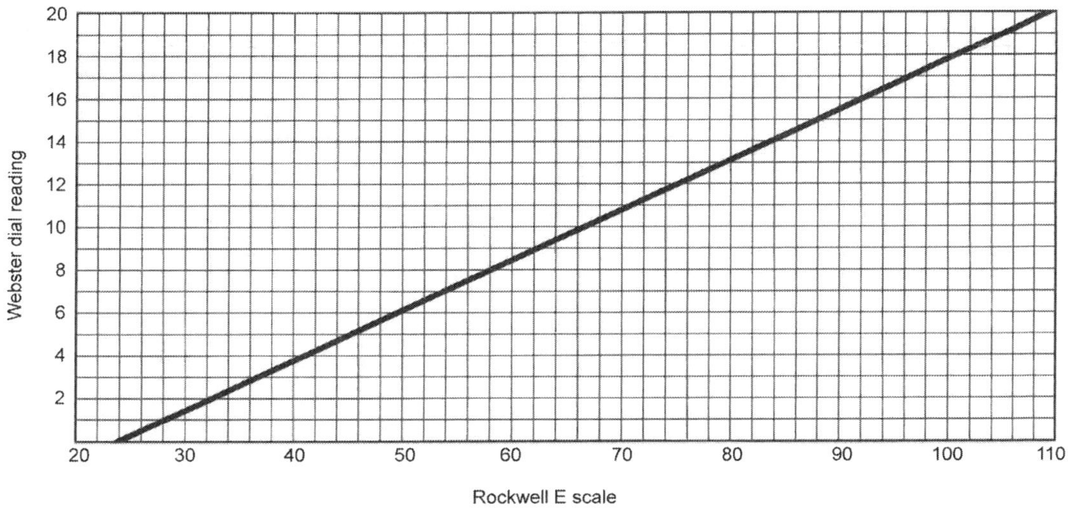

Fig. 4.2 Conversion chart for Webster Model B hardness value to Rockwell E scale. Reprinted with permission of Webster Instruments, Inc

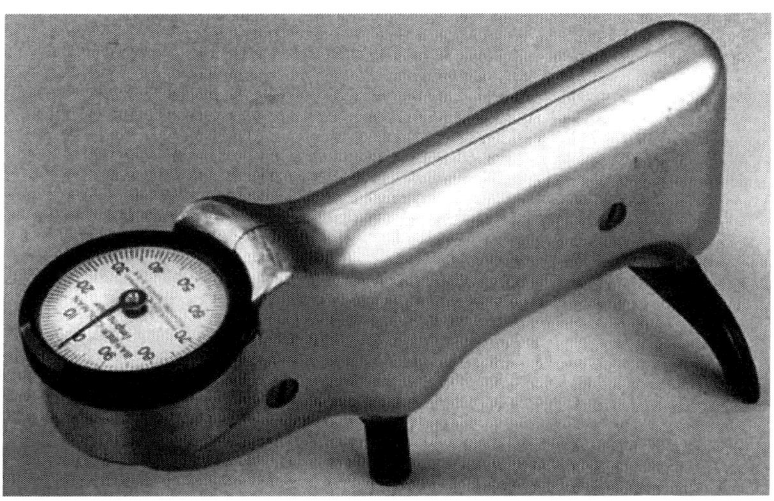

Fig. 4.3 Barcol hardness tester

Relationship to Tensile Strength. However measured, the key to the usefulness of hardness test results to assess the residual strength of aluminum alloy building structures is the relationship established between hardness and ultimate tensile strength. Its usefulness is increased by the fact that within the range of test error for hardness and tensile strength, the relationship is relatively independent of alloy and tempers, although some trends can be identified as illustrated by the following discussion.

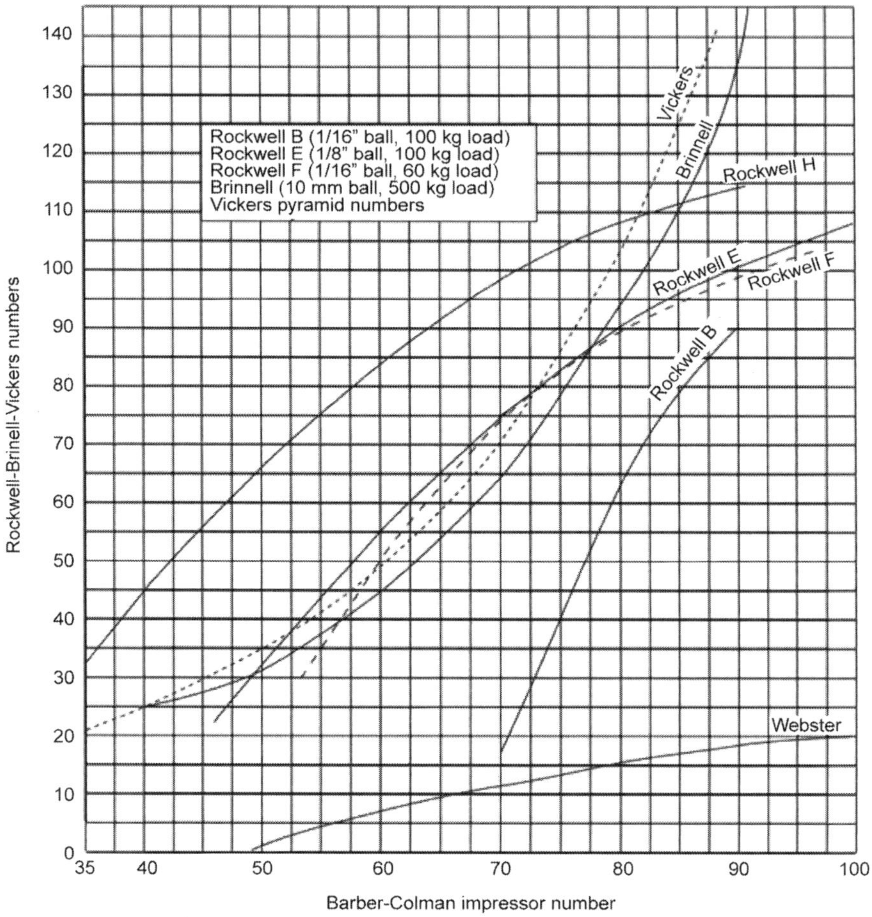

Fig. 4.4 Hardness conversion charts for the Barcol hardness tester

One such relationship that developed over a period of years from tests of hundreds of different lots of aluminum alloy products is illustrated in Fig. 4.7 (Ref 4.1).

The relationship in Fig. 4.7 indicates there is a linear relationship between Brinell hardness (HB) and tensile strength, one that may be characterized as approximately:

$$\text{TS, MPa} = 3.64 \times \text{HB (or, TS, ksi} = 0.528 \times \text{HB)}$$

As is evident in Fig. 4.7, this relationship is not as precise as these calculations might suggest, because the constants in these two equations may vary by as much as ±10%. Some of that variation is contributed by the fact that the upper range of the data, as illustrated in Fig. 4.7, is more likely to represent the non-heat-treatable alloys, while the lower range is more

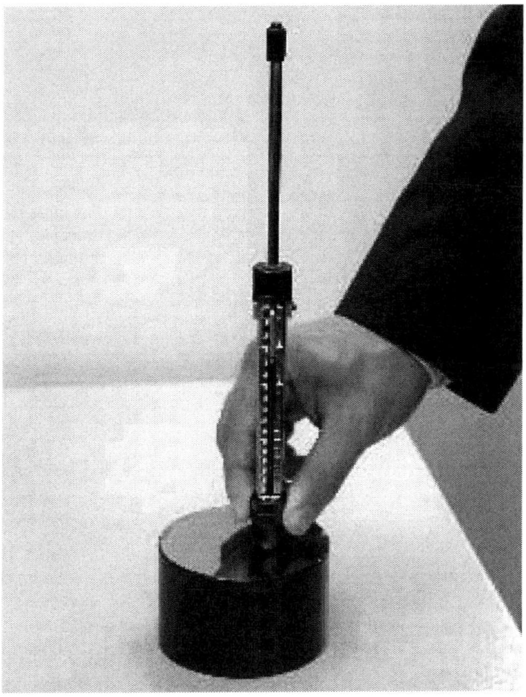

Fig. 4.5 Sclerograph hardness tester

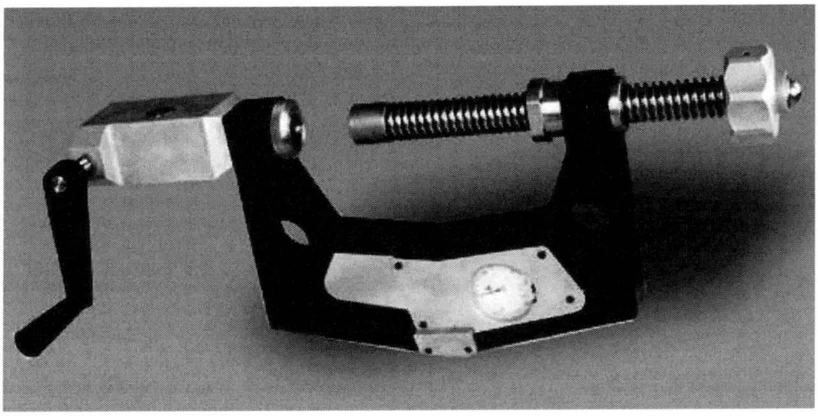

Fig. 4.6 PZ3 Brinell portable hardness testing clamp

likely to be populated by heat treated alloys of the 2*xxx*, 6*xxx*, and 7*xxx* series in the T6, T7, and T8-type tempers.

Making use of these ranges in Fig. 4.7 and of the relationships of the various hardness scales in Fig. 4.4, the values in Table 4.1 and Fig. 4.8 have been generated to provide a direct means of estimation of tensile

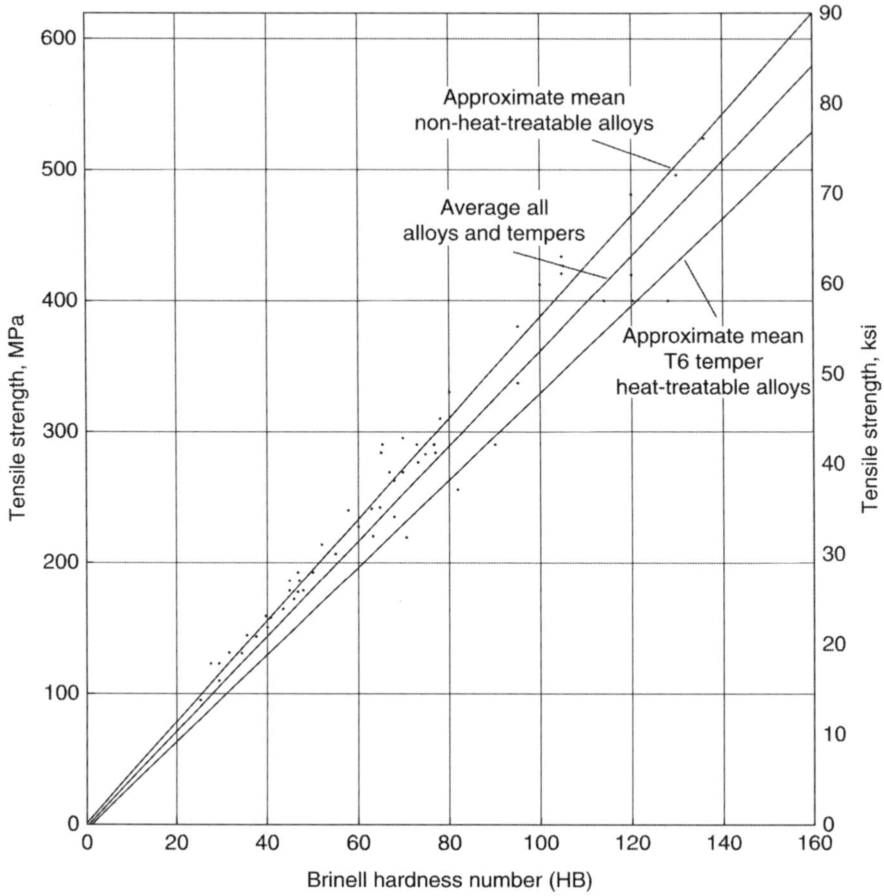

Fig. 4.7 Brinell hardness vs. ultimate tensile strength for aluminum alloys. Source: Ref 4.1

strength of aluminum alloys from the various hardness scales. Emphasis must be placed on the fact that these are estimates of tensile strength, not precise measurements of tensile strength. Nevertheless, they provide a tool for judging the effect of fire exposure on aluminum alloy components, especially if representative values of hardness for the structural members have been determined in advance and stored away in case they would be needed.

Examples of how these relationships and Fig. 4.8 might be used after a fire are:

- A Webster B hardness value of 14.4 has been measured in tests of an extruded 6061-T6 I-beam that was exposed to a fire in a commercial building. From Fig. 4.8, this compares to a tensile strength value of approximately 275 to 315 MPa (40 to 46 ksi), comparable to the typical tensile strength of 6061-T6, 310 MPa (45 ksi), and well above its

Table 4.1 Approximate hardness vs. tensile strength conversions for aluminum alloys

Hardness measurements						Tensile Strength					
Brinell 10 mm², 500 kg	Rockwell			Barcol 934-1	Webster Model B	Average all tempers		O,W,F,T3,T4 tempers		T6, T7, T8 tempers	
	B	E	H			MPa	ksi	MPa	ksi	MPa	ksi
20	...	...	32	35	...	72	11	78	11	66	9.6
21	...	...	33	36	...	76	11	81	12	70	10
22	...	...	37	37	...	80	12	85	12	73	11
23	...	...	40	38	...	83	12	89	13	76	11
24	...	...	43	39	...	87	13	93	14	79	12
25	...	...	45	40	...	91	13	97	14	83	12
26	...	...	47	41	...	94	14	101	15	86	12
26	...	...	49	42	...	94	14	101	15	86	12
27	...	...	52	43	...	98	14	105	15	89	13
27	...	...	54	44	...	98	14	105	15	89	13
27	...	...	56	45	...	98	14	105	15	89	13
28	...	...	58	46	...	101	15	109	16	93	13
29	...	24	61	47	...	105	15	113	16	96	14
30	...	25	63	48	0.7	109	16	116	17	99	14
31	...	28	64	49	1.3	112	16	120	17	103	15
32	...	30	66	50	1.9	116	17	124	18	106	15
33	...	33	68	51	2.5	119	17	128	19	109	16
34	...	36	70	52	3.1	123	18	132	19	113	16
35	...	39	72	53	3.6	127	18	136	20	116	17
37	...	42	73	54	4.2	134	19	144	21	122	18
38	...	44	75	55	4.7	138	20	147	21	126	18
39	...	46	76	56	5.3	141	20	151	22	129	19
40	...	48	78	57	5.8	145	21	155	23	132	19
42	...	51	80	58	6.3	152	22	163	24	139	20
44	...	53	81	59	6.8	159	23	171	25	146	21
45	...	55	83	60	7.3	163	24	175	25	149	22
47	...	57	84	61	7.6	170	25	182	26	156	23
48	...	59	86	62	8.3	174	25	186	27	159	23
50	...	62	88	63	8.8	181	26	194	28	166	24
52	...	64	89	64	9.2	188	27	202	29	172	25
54	...	65	90	64	9.7	196	28	209	30	179	26
55	...	67	91	66	10.1	199	29	213	31	182	26
58	...	69	92	67	10.8	210	30	225	33	192	28
60	...	71	94	68	11	217	32	233	34	199	29
62	...	73	95	69	11.4	224	33	241	35	205	30
64	18	74	96	70	11.8	232	34	248	36	212	31
67	19	76	98	71	12.2	243	35	260	38	222	32
69	28	77	99	72	12.6	250	36	268	39	228	33
71	33	79	100	73	12.9	257	37	275	40	235	34

(continued)

Table 4.1 (continued)

Hardness measurements						Tensile Strength					
Brinell 10 mm², 500 kg	Rockwell B	Rockwell E	Rockwell H	Barcol 934-1	Webster Model B	Average all tempers MPa	Average all tempers ksi	O,W,F,T3,T4 tempers MPa	O,W,F,T3,T4 tempers ksi	T6, T7, T8 tempers MPa	T6, T7, T8 tempers ksi
73	39	81	101	74	13.3	264	38	283	41	242	35
76	45	83	102	75	13.7	275	40	295	43	252	36
80	48	84	103	76	14	290	42	310	45	265	38
84	52	86	104	77	14.3	304	44	326	47	278	40
87	56	88	105	78	14.7	315	46	338	49	288	42
90	60	89	106	79	15	326	47	349	51	298	43
94	63	90	107	80	15.3	340	49	365	53	311	45
97	65	91	108	81	15.6	351	51	376	55	321	47
100	69	92	108	82	15.9	362	53	388	56	331	48
105	72	94	109	83	16.2	380	55	407	59	348	50
109	75	95	109	84	16.4	395	57	423	61	361	52
113	77	96	110	85	16.7	409	59	438	64	374	54
117	80	97	111	86	16.9	424	61	454	66	387	56
122	83	98	111	87	17.2	442	64	473	69	404	59
126	86	99	112	88	17.4	456	66	489	71	417	60
131	89	100	112	89	17.6	474	69	508	74	434	63
135	91	101	113	90	17.8	489	71	524	76	447	65
139	...	102	113	91	18	503	73	539	78	460	67
145	...	103	...	92	18.2	525	76	563	82	480	70
150	...	103	...	93	18.4	543	79	582	84	497	72
155	...	104	...	94	18.6	561	81	601	87	513	74
160	...	104	...	95	18.7	579	84	621	90	530	77

design or minimum tensile strengths, 260 MPa (38 ksi). These results indicate that this particular sample of 6061-T6 could probably continue to be used safely in service.

- A Webster hardness value of 13.4 has been measured on another component of extruded 6061-T6 I-beam exposed to a more severe fire. From Fig. 4.8, this relates to tensile strength values from approximately 225 to 270 MPa (33 to 39 ksi), balancing around but mostly below the design tensile strength of 260 MPa (38 ksi). This suggests that the tensile strength of the 6061-T6 component has been reduced to below its design tensile strength and probably should be taken out of service and replaced.

When the estimated tensile strength values are closer to the minimum or design values, it becomes a matter of judgment as to whether or not a member should be taken out of service, and it may well depend on how highly stressed it is likely to be under static or fatigue loading and whether

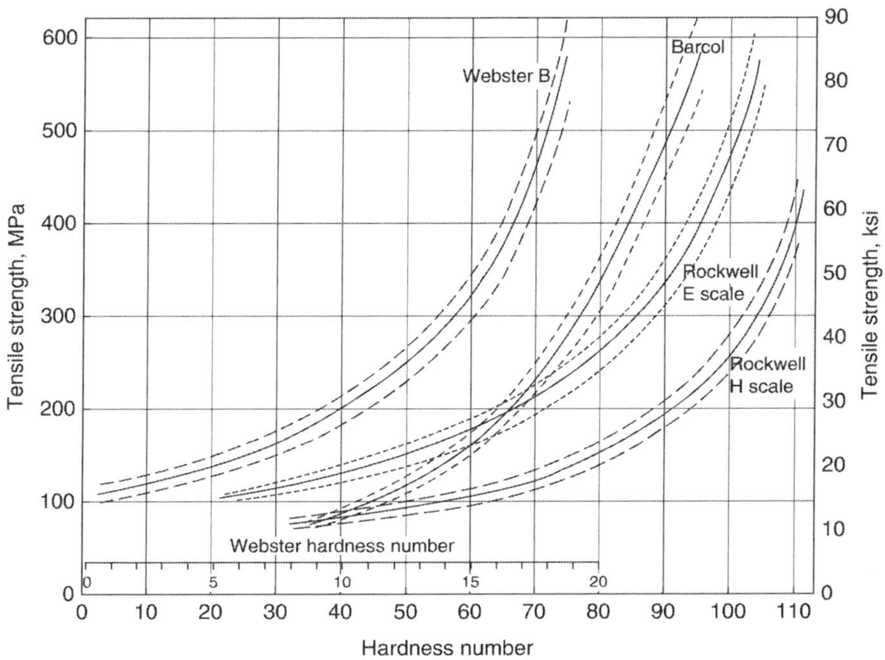

Fig. 4.8 Graphical conversions of hardness and tensile strength of aluminum alloys

or not it is a principal load-carrying member or an auxiliary or redundant member.

Vickers Hardness Measurements. While the previous illustrations have focused primarily on portable hardness testers most easily used in the field, Vickers diamond penetration hardness test measurements (ASTM Standard Test Method E384) may also be useful for such studies where suitable access to the structural elements under consideration can be obtained. As illustrated in Ref 4.2, reasonably good correlations between Vickers hardness measurements and tensile strength may also be obtained.

4.2 Electrical Conductivity Tests

Like hardness measurements, electrical conductivity (EC) measurements may also be used to evaluate the effects of exposure to fire for aluminum alloys. While the relationship between conductivity and strength is more complex than the relationship between hardness and strength, and it varies alloy to alloy and even sometimes by temper, it may still be useful.

An illustration of a relationship between EC and tensile strength comes from a study done by the Aeronautical Materials Laboratory of the Naval Air Engineering Center in 1964 (Ref 4.3). Data from these tests are tabulated in Table 4.2 and summarized in the plots in Fig. 4.9.

Table 4.2 Results of tensile and electrical conductivity tests of some aluminum alloys(a)

Alloy and temper	Sheet thickness, mm(in.)	Exposure temperature, °C (°F)	Exposure time, h	Electrical conductivity, %IACS(b)	Tensile strength		Yield strength	
					MPa	ksi	MPa	ksi
2020-T6	1.6 (0.064)	none	none	20.5	543	78.7	507	73.5
		205 (400)	1	20.5	530	76.8	494	71.6
		205 (400)	5	21.0	510	73.9	469	68.0
		260 (500)	1	21.9	406	S8.8	320	46.4
		315 (600)	1	23.3	318	46.1	194	28.1
2024-H	3.2 (0.125)	none	none	29.0	480	69.6	313	45.4
		205 (400)	1	30.0	463	67.2	297	43.0
		265 (510)	0.17	38.S	448	65.0	387	56.1
		254 (490)	1	38.8	472	68.4	416	60.3
		315 (600)	0.17	40.3	390	56.5	283	41.0
		315 (600)	1	41.5	396	57.4	281	40.8
		315 (600)	3	42.5	313	45.4	183	26.6
		370 (700)	2	42.5	273	39.6	110	15.9
2024-T81	3.2 (0.125)	none	none	38.0	497	72.1	458	66.4
		205 (400)	1	37.5	497	72.0	457	66.3
		205 (400)	5	38.7	488	70.8	440	63.8
		265 (510)	0.17	38.8	463	67.1	390	56.6
		254 (490)	1	39.3	463	67.1	397	57.5
		315 (600)	0.17	40.0	399	57.8	279	40.5
		315 (600)	1	41.0	375	54.4	252	36.6
2219-T81	1.6 (0.062)	none	none	32.0	464	67.3	359	52.1
		205 (400)	1	32.0	474	68.7	359	52.1
		205 (400)	5	32.5	321	46.5	241	34.9
		205 (400)	5	32.5	321	46.6	245	35.5
		260 (500)	1	33.5	408	59.2	281	40.8
		315 (600)	1	34.6	352	51.0	229	33.2
6061-T6	1.6 (0.063)	none	none	39.0	322	46.7	282	40.9
		205 (400)	1	40.0	322	46.7	276	40.0
		205 (400)	5	40.S	303	44.0	281	40.8
		265 (510)	0.17	41.0	264	38.3	228	33.0
		254 (490)	1	41.0	284	41.2	241	35.0
		315 (600)	0.17	42.3	197	28.6	129	18.7
		315 (600)	1	42.5	191	27.7	115	16.7
7002-T6	1.6 (0.062)	none	none	33.5	481	69.7	406	58.9
		205 (400)	0.17	35.0	452	65.5	374	54.2
		205 (400)	1	36.0	461	66.8	373	54.1
		265 (510)	0.17	37.5	346	50.2	204	29.6
		254 (490)	1	38.5	360	52.2	225	32.6

(continued)

(a) Reference:Aeronautical Materials Laboratory Report No.NAEC-AML-2083, "Investigation Into the Electrical Conductivity & Mechanical Properties of Aluminum Alloys Subjected to Elevated Temperature Exposure," by William Allen & Robert G. Mahorter, Naval Air Engineering Center, Philadelphia, Nov. 30,1964. (b) Determined with Magnatest Conductivity Meter, F100 series.

Table 4.2 (continued)

Alloy and temper	Sheet thickness, mm(in.)	Exposure temperature, °C (°F)	Exposure time, h	Electrical conductivity, %IACS(b)	Tensile strength		Yield strength	
					MPa	ksi	MPa	ksi
		315 (600)	0.17	37.0	288	41.8	119	17.3
		315 (600)	1	38.5	284	41.2	114	16.6
7075-T6	2.5 (0.100)	none	none	32.0	590	85.5	545	79.0
		205 (400)	1	37.0	549	79.6	499	72.4
		205 (400)	5	41.0	412	59.7	310	45.0
		265 (510)	0.17	41.5	375	54.3	245	35.5
		254 (490)	1	41.0	390	56.6	270	39.2
		315 (600)	0.17	41.5	305	44.2	138	20.0
		315 (600)	1	42.0	283	41.1	128	18.6
7075-T73	1.7 (0.065)	none	none	37.5	519	75.2	438	63.5
		205 (400)	1	38.5	501	72.6	406	58.8
		205 (400)	5	40.5	410	59.5	302	43.8
		265 (510)	0.17	41.5	366	53.1	239	34.6
		254 (490)	1	41.0	385	55.8	258	37.4
		315 (600)	0.17	39.5	303	43.9	134	19.5
		315 (600)	1	41.5	288	41.7	130	18.9
7079-T6	(0.088)	none	none	31.0	552	80.0	483	70.0
		205 (400)	0.17	33.0	512	74.3	431	62.5
		205 (400)	1	34.0	512	74.2	431	62.5
		205 (400)	5	36.5	424	61.5	308	44.6
		265 (510)	0.17	35.8	390	56.6	239	34.7
		254 (490)	1	36.5	401	58.2	261	37.9
		315 (600)	0.17	36.0	331	48.0	143	21.2
		315 (600)	1	37.0	310	45.0	143	20.8
7178-T6	(0.1)	none	none	29.5	633	91.8	581	84.2
		205 (400)	1	36.0	588	85.2	543	78.8
		205 (400)	5	41.0	433	62.8	331	48.0
		265 (510)	0.17	40.5	386	55.9	258	37.4
		254 (490)	1	41.3	405	58.7	288	41.8
		315 (600)	0.17	40.5	314	45.5	149	21.6
		315 (600)	1	41.8	291	42.2	138	20.0

(a) Reference:Aeronautical Materials Laboratory Report No.NAEC-AML-2083, "Investigation Into the Electrical Conductivity & Mechanical Properties of Aluminum Alloys Subjected to Elevated Temperature Exposure," by William Allen & Robert G. Mahorter, Naval Air Engineering Center, Philadelphia, Nov. 30,1964. (b) Determined with Magnatest Conductivity Meter, F100 series.

Figure 4.9 illustrates that there are also potentially useful relationships between EC and tensile strength, and that the relationship is alloy and temper specific. In most cases, an increase in EC measurements will indicate some loss in strength as a result of the exposure. However, the slopes or gradients of the relationships for most alloys are relatively small with respect to the change in EC associated with change in tensile strength, which limits their usefulness. And for some alloys, for example, the higher

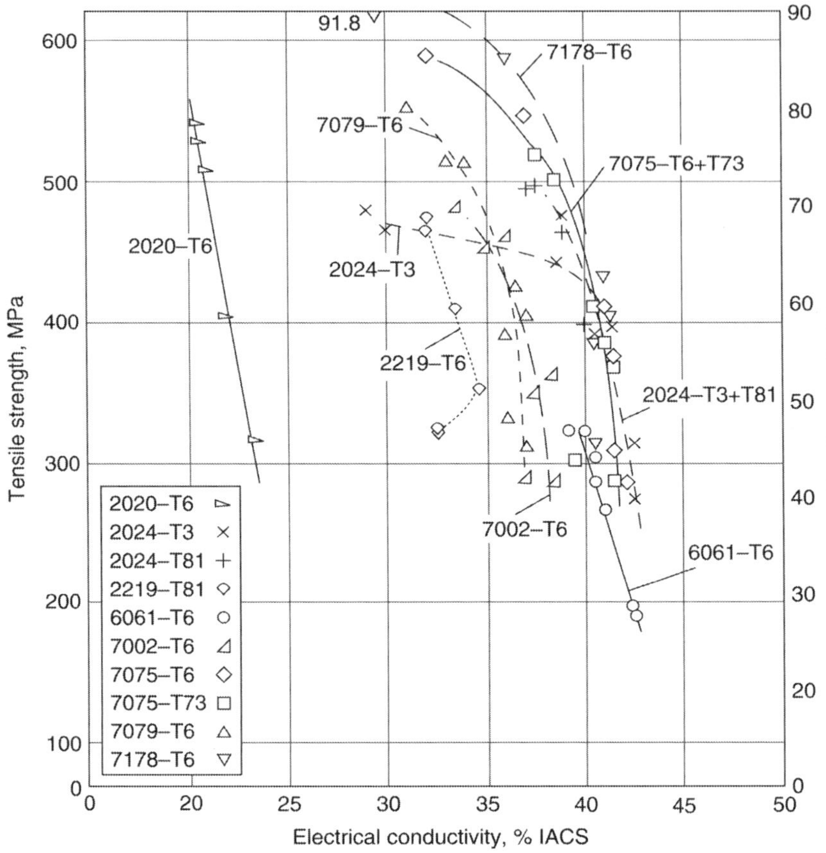

Fig. 4.9 Electrical conductivity vs. tensile strength for some aluminum alloys. Source: Ref 4.3

EC values for 2024 and for 2219-T6, the gradient is so small as to render the relationships not very useful.

The results of an Aluminum Association study of the use of EC measurements in comparison with tensile properties are shown in Fig. 4.10 (Ref 4.4). Data generated for 2124-T851 plate indicate a relatively useful and discriminating relationship between EC and tensile strength.

The specified minimum or design tensile strength of relatively thin 2124-T851 sheet and plate is 455 MPa (66 ksi); using Fig. 4.10, if EC test results for a 2124-T851 structural member fall above approximately 41% IACS (International Annealed Copper Standard), the usefulness of that member may be considered questionable. Unfortunately, there are not very many such useful test-data-based relationships for other alloys and tempers.

Overall, these data suggest that EC is not likely to be as useful as hardness testing to measure any loss in strength associated with fire exposure, but EC measurements might be helpful in combination with the hardness

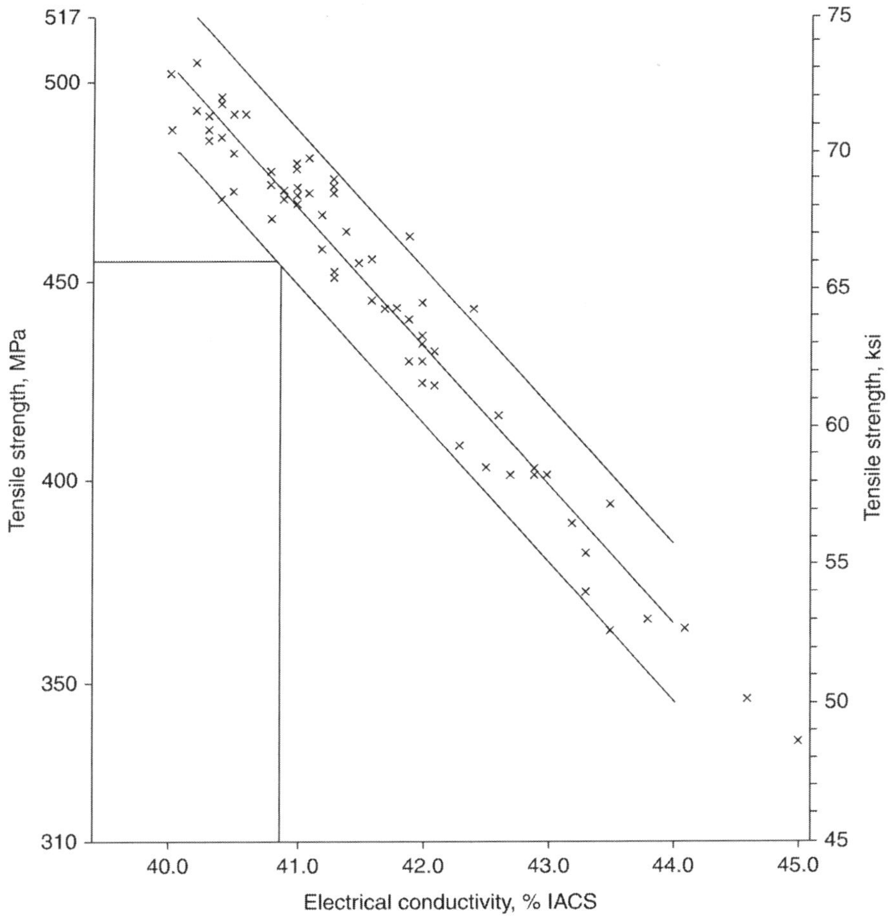

Fig. 4.10 Electrical conductivity vs. tensile strength for 2124-T851. Source: Ref 4.4

tests to pin down important losses. The findings of other investigators (Ref 4.5) on aluminum alloy 7010 were consistent with these conclusions.

4.3 Summary of Findings Regarding Estimate of Fire Damage

Based on the investigations discussed herein, it appears that measurements of hardness after fire exposure may be expected to give the most reliable measure of fire damage resulting from the high-temperature exposure, especially if benchmark values for typical members have been obtained beforehand. In all such studies, it will be important to use hardness measuring equipment and procedures appropriate for the thicknesses of the structural members that are the focus of the study. It is important to avoid potential secondary effects on hardness test results such as those resulting from the anvil effect when relatively thin members are tested.

Electrical conductivity tests by themselves do not appear to be as consistently and reliably useful as hardness tests in measuring fire damage because of (a) the relatively small gradient in EC associated with significant loss in strength for some alloys and tempers, and (b) the relatively few reliable relationships available for different alloys and tempers. Electrical conductivity may be a useful secondary measurement together with hardness in increasing the conclusiveness of such studies if supporting data are available.

REFERENCES

4.1 Unpublished data from Alcoa via The Aluminum Association, 1952

4.2 R.D. Matulich, "Post-fire Mechanical Properties of Aluminum Alloys and Aluminum Welds," Master of Science Thesis, Virginia Polytechnic Institute and State University, Blacksburg, VA, April 22, 2011

4.3 W. Allen and R.S. Mahorter, "Investigation into the Electrical Conductivity and Mechanical Properties of Aluminum Alloys Subjected to Elevated Temperature Exposure," Aeronautical Materials Laboratory Report No. NAEC-AML-2083, Naval Air Engineering Center, Philadelphia, PA, Nov 30, 1964

4.4 "Report on Electrical Conductivity in Heat Treated Aluminum Alloy Plate," The Aluminum Association, Washington, D.C., June 1, 1991

4.5 M.A. Salazar-Guapuriche, Y.Y. Zhao, A. Pitman, and A. Greene, Correlation of Strength and Electrical Conductivity for Aluminium Alloy 7010, *Mat. Sci. Forum,* Vol 519–521, 2006, p 853–858

CHAPTER 5

Applications Not Recommended for Aluminum Alloys

THE RELATIVELY LOW MELTING RANGES for aluminum and its alloys (Table 1.1 and Appendix 2) in comparison with the temperatures likely to be reached in significant conflagrations suggest that these materials not be used in applications where exposure to temperatures above approximately 475 °C (890 °F) would be anticipated.

Further, in consideration of the significant reduction in strength of most aluminum alloys at temperatures above 150 to 200 °C (300 to 400 °F), illustrated in Appendix 1, their continuous use in applications where these temperatures are exceeded should be avoided. In this regard, aluminum alloys of the aluminum-copper (2*xxx*) series have been designed specifically for use at moderately elevated temperatures, so they may be considered for carefully designed applications up to approximately 315 °C (600 °F).

Examples of unsuitable applications for aluminum and aluminum alloys include:

- Furnaces for the melting of metals and alloys and processing recycled scrap, or any other oven or furnace operation operating continuously above 200 °C (approximately 400 °F);
- Motor and engine components (such as rotors) operating continuously above 200 °C;
- Fire doors, although they would operate as a significant heat sink to allow added time for individuals to exit the structure in peril.

CHAPTER 6

Summary

ALUMINUM melts at approximately 655 °C (1200 °F) and experiences significant loss in strength at temperatures above approximately 200 °C (400 °F). In an engulfing fire, it will melt and run off, but aluminum does not burn in air nor will it support combustion.

When aluminum and aluminum alloys are tested in accordance with ASTM Standards E108 or E136, British Standard 476, or building codes such as that for the city of Los Angeles, aluminum has achieved the highest ratings for being not easily ignitable and received the highest classifications for resistance to flame spread and fire penetration. An excellent summary of information on this subject is contained on the Aluminum Federation website and in Appendix 5.

Aluminum alloy structures can be protected against fire for suitable periods of time as selected by various specifying authorities using commercial fireproofing technologies such as lightweight vermiculite concrete coating or rock wool insulation.

The high reflectivity, thermal conductivity, and specific heat of aluminum aid in providing resistance to structures against temperature rise during a fire comparable to that provided by steels.

Aluminum has been successfully used for many years in applications requiring high resistance to burning and to flame spread, including oil rigs, ships, and flammable-fluid-hauling trailer trucks.

Despite erroneous press reports, there is no evidence that aluminum contributed to the loss of ships such as the HMS *Sheffield* (an all-steel ship) in the Falklands Islands War, as determined by the British Ministry of Defense, nor to the damage to the USS *Stark* in the Persian Gulf confrontation with Iraq, as stated by U.S. Navy spokesmen.

Estimates may be made of the significance of damage to aluminum member exposed to fire using portable hardness testers and existing correlations relating hardness and tensile strength. Electrical conductivity measurements may also be useful in such studies if sufficient supporting data are available.

Supporting information regarding the excellent fire resistance of aluminum and aluminum alloys is contained in Ref 6.1 through 6.6 and on websites such as that of the Aluminium Federation (Ref 6.7).

REFERENCES

6.1 J.A. Purkiss and L.-Y. Li, *Fire Safety Engineering Design of Structures,* 3rd ed., CRC Press, New York, 2013

6.2 B. Faggiano, G. De Matteis, R. Landolfo, and F.M. Mazzolani, Behaviour of Aluminium Structures Under Fire, *J. Civ. Eng. Manag.,* Vol X (No. 3), 2002, p 183–190

6.3 M.J. Bayley, The Fire Protection of Aluminium in Offshore Structures, *Proceedings of the International Mechanical Engineering (IMechE) Conference on Materials and Design against Fire,* Mechanical Engineering Publications, London, 1992, p 113–120

6.4 S. Lundberg, "Material Aspects of Fire Design," TALAT Lecture 2502, European Aluminium Association, 1994

6.5 "Fire Resistance and Flame Spread Performance of Aluminum and Aluminum Alloys," Standard AA FRFS, 2nd ed., The Aluminum Association, Washington, D.C., July 2002

6.6 Fire Resistance of Aluminum, *Aluminum and the Sea,* Alcan Aluminium Company, 2013

6.7 "Aluminium and Fire," UK Aluminium Industry Fact Sheet 11, ALFED (Aluminium Federation Ltd), http://www.alfed.org.uk/files/Fact%20sheets/11-aluminium-and-fire.pdf (accessed Jan 7, 2016)

APPENDIX 1

Elevated Temperature Tensile Properties of Representative Alloys*

1100–O: Typical tensile properties

Temperature		Time at temperature, h	Tensile strength		Yield strength		Elongation in 4*D*, %	Modulus of elasticity(a)	
°F	°C		ksi	MPa	ksi	MPa		10^6 psi	GPa
−452	−269	...	46	315	8.4	58	37		
−320	−196	...	25	170	6.0	41	55	11.1	77
−112	−80	...	15	105	5.5	38	43	10.4	72
−18	−28	...	14	95	5.0	34	40	10.1	70
75	25	...	13	90	5.0	34	40	9.9	68
212	100	0.5	11	75	4.6	32	45	...	...
		10	11	75	4.6	32	45	...	...
		100	11	75	4.6	32	45	...	...
		1,000	11	75	4.6	32	45	...	...
		10,000	11	75	4.6	32	45	...	...
300	150	0.5	8.5	59	4.2	29	55	...	...
		10	8.5	59	4.2	29	55	...	...
		100	8.5	59	4.2	29	55	...	...
		1,000	8.5	59	4.2	29	55	...	...

(continued)

(a) The modulus of elasticity in compression is about 2% greater than in tension. Source data are in English units; metric values are converted and rounded.

*Reprinted from J.G. Kaufman, *Properties of Aluminum Alloys: Tensile, Creep, and Fatigue Data at High and Low Temperatures,* The Aluminum Association and ASM International, 1999

1100–O (continued)

Temperature		Time at temperature, h	Tensile strength		Yield strength		Elongation in 4*D*, %	Modulus of elasticity(a)	
°F	°C		ksi	MPa	ksi	MPa		10^6 psi	GPa
		10,000	8.5	59	4.2	29	55	...	...
350	177	0.5	7.5	52	3.8	26	60	...	...
		10	7.5	52	3.8	26	60	...	...
		100	7.5	52	3.8	26	60	...	...
		1,000	7.5	52	3.8	26	60	...	...
		10,000	7.5	52	3.8	26	60	...	...
400	205	0.5	6.0	41	3.5	24	65	...	...
		10	6.0	41	3.5	24	65	...	...
		100	6.0	41	3.5	24	65	...	...
		1,000	6.0	41	3.5	24	65	...	...
		10,000	6.0	41	3.5	24	65	...	...
450	230	0.5	5.0	34	3.1	21	70	...	...
		10	5.0	34	3.1	21	70	...	...
		100	5.0	34	3.1	21	70	...	...
		1,000	5.0	34	3.1	21	70	...	...
		10,000	5.0	34	3.1	21	70	...	...
500	260	0.5	4	28	2.6	18	75	...	...
		10	4	28	2.6	18	75	...	...
		100	4	28	2.6	18	75	...	...
		1,000	4	28	2.6	18	75	...	...
		10,000	4	28	2.6	18	75	...	...
600	315	0.5	2.9	20	2	14	80	...	...
		10	2.9	20	2	14	80	...	...
		100	2.9	20	2	14	80	...	...
		1,000	2.9	20	2	14	80	...	...
		10,000	2.9	20	2	14	80	...	...
700	370	0.5	2.1	14	1.6	11	85	...	...
		10	2.1	14	1.6	11	85	...	...
		100	2.1	14	1.6	11	85	...	...
		1,000	2.1	14	1.6	11	85	...	...
		10,000	2.1	14	1.6	11	85	...	...

(a) The modulus of elasticity in compression is about 2% greater than in tension. Source data are in English units; metric values are converted and rounded.

2014–1651 Plate 0.250 to 2.000 in. (>6.30 ≤ 50.00 mm) thick: typical tensile properties

Temperature		Time at temperature, h	Tensile strength		Yield strength		Elongation in 4*D*, %	Modulus of elasticity(a)	
°F	°C		ksi	MPa	ksi	MPa		10^6 psi	GPa
−452	−269	...	97	670	79	545	15	...	...
−423	−253	...	97	670	77	530	16	...	...
−320	−196	...	84	580	70	485	14	11.8	81
−112	−80	...	74	510	63	435	12	11.0	76
−18	−28	...	72	495	62	425	12	10.7	74
75	25	...	70	485	60	415	13	10.5	72
212	100	0.5	63	435	57	395	14	10.3	71
		10	63	435	57	395	14	10.3	71
		100	63	435	58	400	14	10.3	71
		1,000	64	440	58	400	15	10.3	71
		10,000	64	440	59	405	15	10.3	71
		100,000	61	420	55	380	11	10.3	71
300	150	0.5	55	380	51	350	15	9.9	68
		10	56	385	51	350	16	9.9	68
		100	56	385	51	350	16	9.9	68
		1,000	51	350	46	315	17	9.9	68
		10,000	40	275	35	240	20	9.9	68
		100,000	25	170	19	130	30	9.9	68
350	177	0.5	51	350	47	325	14	9.6	66
		10	50	345	46	315	17	9.6	66
		100	45	310	41	285	18	9.6	66
		1,000	34	235	30	205	20	9.6	66
		10,000	25	170	20	140	28	9.6	66
		100,000	19	130	14	95	38	9.6	66
400	205	0.5	45	310	41	285	14	9.2	63
		10	41	285	37	255	18	9.2	63
		100	30	205	27	185	22	9.2	63
		1,000	21	145	18	125	29	9.2	63
		10,000	16	110	13	90	38	9.2	63
		100,000	15	105	11	75	46	9.2	63
450	230	0.5	37	255	33	230	15	8.9	61
		10	28	195	26	180	22	8.9	61
		100	20	140	17	115	29	8.9	61
		1,000	15	105	12	85	36	8.9	61
		10,000	12	85	9.5	66	45	8.9	61
		100,000	11	75	9.0	62	54	8.9	61
500	260	0.5	25	170	23	160	18	8.5	59
		10	16	110	15	105	27	8.5	59

(continued)

(a) The modulus of elasticity in compression is about 2% greater than in tension. Source data are in English units; metric values are converted and rounded.

2014–1651 Plate 0.250 to 2.000 in. (>6.30 ≤ 50.00 mm) thick (continued)

Temperature		Time at temperature, h	Tensile strength		Yield strength		Elongation in 4*D*, %	Modulus of elasticity(a)	
°F	°C		ksi	MPa	ksi	MPa		10^6 psi	GPa
		100	13	90	11	75	34	8.5	59
		1,000	11	75	9.5	66	43	8.5	59
		10,000	9.5	66	7.5	52	52	8.5	59
		100,000	9.0	62	7.5	52	60	8.5	59
600	315	0.5	11	75	9.5	66	28	7.7	53
		10	9.0	62	7.0	48	39	7.7	53
		100	7.5	52	6.0	41	48	7.7	53
		1,000	7.0	48	5.5	38	55	7.7	53
		10,000	6.5	45	5.0	34	65	7.7	53
		100,000	6.5	45	5.0	34	72	7.7	53
700	370	0.5	6.0	41	5.0	34	50	6.5	45
		10	5.0	34	4.2	29	56	6.5	45
		100	4.8	33	3.8	26	62	6.5	45
		1,000	4.5	31	3.7	26	68	6.5	45
		10,000	4.3	30	3.5	24	72	6.5	45
		100,000	4.3	30	3.5	24	75	6.5	45
800	425	0.5	3.7	26	2.7	19	67	...	...
900	480	0.5	2.2	15	1.6	11	77	...	...
1000	540	0.5	1.2	8.0	0.7	5.0	14	...	...

(a) The modulus of elasticity in compression is about 2% greater than in tension. Source data are in English units; metric values are converted and rounded.

2024-T3: Typical tensile properties

Temperature		Time at temperature, h	Tensile strength		Yield strength		Elongation in 4*D*, %	Modulus of elasticity(a)	
°F	°C		ksi	MPa	ksi	MPa		10^6 psi	GPa
−423	−253	...	112	770	75	515	17	...	...
−320	−196	...	85	585	62	425	18	11.8	81
−112	−80	...	73	*505*	52	360	17	11.0	76
−18	−28	...	72	495	51	350	17	10.7	74
75	25	...	70	485	50	345	17	10.5	72
212	100	0.1	66	455	48	330	17	10.3	71
		0.5	66	455	48	330	17	10.3	71
		10	66	455	48	330	17	10.3	71
		100	66	455	48	330	17	10.3	71
		1,000	66	455	49	340	17	10.3	71
		10,000	67	460	51	350	16	10.3	71

(continued)

(a) The modulus of elasticity in compression is about 2% greater than in tension. Source data are in English units; metric values are converted and rounded.

2024-T3 (continued)

Temperature		Time at temperature, h	Tensile strength		Yield strength		Elongation in 4*D*, %	Modulus of elasticity(a)	
°F	°C		ksi	MPa	ksi	MPa		10^6 psi	GPa
300	150	0.1	...	...	...	...	...	9.9	68
		0.5	60	415	45	310	18	9.9	68
		10	60	415	46	315	17	9.9	68
		100	63	435	53	365	15	9.9	68
		1,000	58	400	55	380	11	9.9	68
		10,000	54	370	49	340	13	9.9	68
		100,000	50	345	43	295	15	9.9	68
350	177	0.5	57	395	43	295	14	9.6	66
		10	59	405	54	370	13	9.6	66
		100	54	370	49	340	12	9.6	66
		1,000	48	330	44	305	14	9.6	66
		10,000	43	295	36	250	16	9.6	66
		100,000	32	220	26	180	23	9.6	66
400	205	0.5	54	370	48	330	13	9.2	63
		10	48	330	45	310	14	9.2	63
		100	44	305	39	270	15	9.2	63
		1,000	38	260	32	220	19	9.2	63
		10,000	27	185	20	140	28	9.2	63
		100,000	18	125	13	90	40	9.2	63
450	230	0.5	45	310	41	285	15	8.9	61
		10	40	275	36	250	17	8.9	61
		100	34	235	29	200	20	8.9	61
		1,000	25	170	18	125	30	8.9	61
		10,000	16	110	11	75	45	8.9	61
		100,000	12	85	8.5	59	55	8.9	61
500	260	0.1	39	270	35	240	17	8.5	59
		0.5	37	255	33	230	17	8.5	59
		10	30	205	27	185	20	8.5	59
		100	22	150	18	125	29	8.5	59
		1,000	15	105	11	75	45	8.5	59
		10,000	11	75	8.0	55	60	8.5	59
		100,000	9.0	62	6.5	45	65	8.5	59
600	315	0.1	23	160	21	145	23	7.7	53
		0.5	20	140	17	115	26	7.7	53
		10	12	85	10	70	40	7.7	53
		100	10	70	8.0	55	50	7.7	53
		1,000	9.0	62	6.5	45	65	7.7	53
		10,000	7.5	52	6.0	41	75	7.7	53
		100,000	6.5	45	5.5	38	80	7.7	53

(continued)

(a) The modulus of elasticity in compression is about 2% greater than in tension. Source data are in English units; metric values are converted and rounded.

2024-T3 (continued)

Temperature		Time at temperature, h	Tensile strength		Yield strength		Elongation in 4*D*, %	Modulus of elasticity(a)	
°F	°C		ksi	MPa	ksi	MPa		10^6 psi	GPa
700	370	0.1	11	75	10	70	35	6.5	45
		0.5	8.5	59	6.5	45	50	6.5	45
		10	7.0	48	5.0	34	75	6.5	45
		100	6.0	41	4.5	31	85	6.5	45
		1,000	5.5	38	4.1	28	90	6.5	45
		10,000	5.0	34	4.1	28	95	6.5	45
		100,000	4.9	34	4.1	28	100	6.5	45
800	425	0.1	5.0	34	4.1	28	65	...	...
		0.5	4.4	30	3.5	24	85	...	...
900	480	...	2.3	16	1.8	12	65	...	...
1000	540	...	0.3	2.0	0.3	2.0	2	...	...

(a) The modulus of elasticity in compression is about 2% greater than in tension. Source data are in English units; metric values are converted and rounded.

2219–T851 Plate: typical tensile properties

Temperature		Time at temperature, h	Tensile strength		Yield strength		Elongation in 4*D*, %	Modulus of elasticity(a)	
°F	°C		ksi	MPa	ksi	MPa		10^6 psi	GPa
−452	−269	...	96	662	66	455	16	...	...
−423	−253	...	96	662	65	450	17	12.4	85
−320	−196	...	83	572	60	415	14	11.8	81
−112	−80	...	71	490	54	370	12	11.0	76
−18	−28	...	69	475	52	360	12	10.7	74
75	25	...	66	455	50	345	12	10.5	72
212	100	0.1	60	415	47	325	15	10.2	70
		0.5	60	415	47	325	15	10.2	70
		10	60	415	47	325	15	10.2	70
		100	60	415	47	325	15	10.2	70
		1,000	60	415	47	325	15	10.2	70
		10,000	60	415	47	325	15	10.2	70
300	150	0.1	54	370	44	305	17	9.8	68
		0.5	54	370	44	305	17	9.8	68
		10	54	370	44	305	17	9.8	68
		100	54	370	44	305	17	9.8	68
		1,000	52	360	43	295	17	9.8	68
		10,000	49	340	40	275	17	9.8	68
350	177	0.1	...	...	...	...	...	9.4	65
		0.5	49	340	41	285	19	9.4	65

(continued)

(a) The modulus of elasticity in compression is about 2% greater than in tension. Source data are in English units; metric values are converted and rounded.

2219–T851 Plate (continued)

Temperature		Time at temperature, h	Tensile strength		Yield strength		Elongation in 4D, %	Modulus of elasticity(a)	
°F	°C		ksi	MPa	ksi	MPa		10^6 psi	GPa
		10	48	330	40	275	19	9.4	65
		100	45	310	38	260	19	9.4	65
		1,000	43	295	35	240	19	9.4	65
		10,000	41	285	33	230	19	9.4	65
400	205	0.5	43	295	36	250	20	9.1	63
		10	40	275	34	235	20	8.1	56
		100	38	260	32	220	20	9.1	63
		1,000	36	250	30	205	20	8.1	56
		10,000	36	250	29	200	20	9.1	63
450	230	0.5	35	240	30	205	21	8.8	61
		10	34	235	28	195	21	8.8	61
		100	33	230	27	185	21	8.8	61
		1,000	32	220	27	185	21	8.8	61
		10,000	32	220	26	180	21	8.8	61
500	260	0.5	29	200	25	170	21	8.5	59
		10	29	200	24	165	21	8.5	59
		100	29	200	24	165	21	8.5	59
		1,000	29	200	24	165	21	8.5	59
		10,000	29	200	23	160	21	8.5	59
600	315	0.5	20	140	18	125	21	7.6	52
		10	19	130	17	115	21	7.6	52
		100	18	125	15	105	22	7.6	52
		1,000	14	95	12	85	28	7.6	52
		10,000	7.0	48	6.0	41	55	7.6	52
700	370	0.5	11	75	9.0	62	23	6.3	43
		10	6.5	45	5.5	38	40	6.3	43
		100	5.0	34	4.2	29	70	6.3	43
		1,000	4.7	32	3.7	26	75	6.3	43
		10,000	4.4	30	3.7	26	75	6.3	43

(a) The modulus of elasticity in compression is about 2% greater than in tension. Source data are in English units; metric values are converted and rounded.

2618–T651 Plate: typical tensile properties

Temperature		Time at temperature, h	Tensile strength		Yield strength		Elongation in 4*D*, %	Modulus of elasticity(a)	
°F	°C		ksi	MPa	ksi	MPa		10^6 psi	GPa
−452	−269	...	89	614	69	475	14	...	...
−320	−196	...	80	552	65	450	12	...	...
−112	−80	...	70	485	59	405	10	...	...
−18	−28	...	67	460	57	395	10	...	...
75	25	...	64	440	54	370	10	10.7	74
212	100	0.5	60	415	52	360	10	10.4	72
		10	60	415	52	360	10	10.4	72
		100	60	415	52	360	10	10.4	72
		1,000	60	415	52	360	10	10.4	72
		10,000	60	415	52	360	10	10.4	72
300	150	0.5	54	370	49	340	12	10.1	70
		10	54	370	49	340	12	10.1	70
		100	54	370	49	340	12	10.1	70
		1,000	53	365	48	330	12	10.1	70
		10,000	50	345	46	315	13	10.1	70
350	177	0.5	50	345	47	325	13	9.9	68
		10	49	340	46	315	13	9.9	68
		100	48	330	45	310	14	9.9	68
		1,000	45	310	42	290	15	9.9	68
		10,000	42	290	38	260	15	9.9	68
400	205	0.5	45	310	43	295	15	9.7	67
		10	43	295	41	285	15	9.7	67
		100	41	285	39	270	16	9.7	67
		1,000	38	260	35	240	20	9.7	67
		10,000	33	230	29	200	24	9.7	67
450	230	0.5	40	275	38	260	16	9.5	66
		10	37	255	35	240	18	9.5	66
		100	33	230	31	215	20	9.5	66
		1,000	28	195	26	180	28	9.5	66
		10,000	24	165	20	140	34	9.5	66
500	260	0.5	34	235	32	220	20	9.3	64
		10	30	205	28	195	22	9.3	64
		100	25	170	22	150	26	9.3	64
		1,000	20	140	17	115	40	9.3	64
		10,000	14	95	10	70	50	9.3	64
600	315	0.5	23	160	20	140	24	8.7	60
		10	17	115	14	95	40	8.7	60

(continued)

(a) The modulus of elasticity in compression is about 2% greater than in tension. Source data are in English units; metric values are converted and rounded.

2618–T651 Plate (continued)

Temperature		Time at temperature, h	Tensile strength		Yield strength		Elongation in 4*D*, %	Modulus of elasticity(a)	
°F	°C		ksi	MPa	ksi	MPa		10^6 psi	GPa
		100	11	75	8.0	55	66	8.7	60
		1,000	8.0	55	6.0	41	80	8.7	60
		10,000	7.5	52	5.5	38	85	8.7	60
700	370	0.5	11	75	9.0	62	40	7.3	50
		10	7.0	48	5.0	34	95	7.3	50
		100	6.0	41	4.0	28	105	7.3	50
		1,000	5.0	34	3.5	24	110	7.3	50
		10,000	5.0	34	3.5	24	120	7.3	50

(a) The modulus of elasticity in compression is about 2% greater than in tension. Source data are in English units; metric values are converted and rounded.

3003–O Rolled and drawn rod: typical tensile properties

Temperature		Time at temperature, h	Tensile strength		Yield strength		Elongation in 4D, %	Modulus of elasticity(a)	
°F	°C		ksi	MPa	ksi	MPa		10^6 psi	GPa
−320	−196	...	32	220	8.5	59	46	11.1	77
−112	−80	...	19	130	6.5	45	44	10.4	72
−18	−28	...	17	115	6.0	40	42	10.1	70
75	25	...	16	110	6.0	40	40	9.9	68
212	100	...	14	95	6.0	40	42	9.5	66
		0.5	14	95	6.0	40	42	9.5	66
		10	14	95	6.0	40	42	9.5	66
		100	14	95	6.0	40	42	9.5	66
		1,000	14	95	6.0	40	42	9.5	66
		10,000	14	95	6.0	40	42	9.5	66
		100,000	14	95	6.0	40	42	9.5	66
300	150	0.1	11	75	5.5	38	50	9.0	62
		0.5	11	75	5.5	38	50	9.0	62
		10	11	75	5.5	38	50	9.0	62
		100	11	75	5.5	38	50	9.0	62
		1,000	11	75	5.5	38	50	9.0	62
		10,000	11	75	5.5	38	50	9.0	62
		100,000	11	75	5.5	38	50	9.0	62
350	177	0.1	9.5	66	5.0	34	55	8.6	59
		0.5	9.5.	66	5.0	34	55	8.6	59
		10	9.5	66	5.0	34	55	8.6	59
		100	9.5	66	5.0	34	55	8.6	59
		1,000	9.5	66	5.0	34	55	8.6	59

(continued)

(a) The modulus of elasticity in compression is about 2% greater than in tension. Source data are in English units; metric values are converted and rounded.

3003–O Rolled and drawn rod (continued)

Temperature		Time at temperature, h	Tensile strength		Yield strength		Elongation in 4D, %	Modulus of elasticity(a)	
°F	°C		ksi	MPa	ksi	MPa		10^6 psi	GPa
		10,000	9.5	66	5.0	34	55	8.6	59
		100,000	9.5	66	5.0	34	55	8.6	59
400	205	0.1	8.0	55	4.4	30	60	8.2	57
		0.5	8.0	55	4.4	30	60	8.2	57
		10	8.0	55	4.4	30	60	8.2	57
		100	8.0	55	4.4	30	60	8.2	57
		1,000	8.0	55	4.4	30	60	8.2	57
		10,000	8.0	55	4.4	30	60	8.2	57
		100,000	8	55	4.4	30	60	8.2	57
450	230	0.1	6.5	45	3.9	27	65	7.7	53
		0.5	6.5	45	3.9	27	65	7.7	53
		10	6.5	45	3.9	27	65	7.7	53
		100	6.5	45	3.9	27	65	7.7	53
		1,000	6.5	45	3.9	27	65	7.7	53
		10,000	6.5	45	3.9	27	65	7.7	53
		100,000	6.5	45	3.9	27	65	7.7	53
500	260	0.1	5.0	34	3.4	23	70	7.1	49
		0.5	5.0	34	3.4	23	70	7.1	49
		10	5.0	34	3.4	23	70	7.1	49
		100	5.0	34	3.4	23	70	7.1	49
		1,000	5.0	34	3.4	23	70	7.1	49
		10,000	5.0	34	3.4	23	70	7.1	49
		100,000	5.0	34	3.4	23	70	7.1	49
600	315	0.1	3.7	26	2.7	19	70	6.1	42
		0.5	3.7	26	2.7	19	70	6.1	42
		10	3.7	26	2.7	19	70	6.1	42
		100	3.7	26	2.7	19	70	6.1	42
		1,000	3.7	26	2.7	19	70	6.1	42
		10,000	5.7	39	2.7	19	70	6.1	42
		100,000	3.7	26	2.7	19	70	6.1	42
700	370	0.1	3.0	21	2.2	15	70	5.0	34
		0.5	3.0	21	2.2	15	70	5.0	34
		10	3.0	21	2.2	15	70	5.0	34
		100	3.0	21	2.2	15	70	5.0	34
		1,000	3.0	21	2.2	15	70	5.0	34
		10,000	3.0	21	2.2	15	70	5.0	34
		100,000	3.0	21	2.2	15	70	5.0	34
800	425	...	2.1	14	1.7	12	70	...	...
900	480	...	1.6	11	1.2	8.0	70	...	...
1000	540	...	1.4	10	0.9	6.0	70	...	...

(a) The modulus of elasticity in compression is about 2% greater than in tension. Source data are in English units; metric values are converted and rounded.

3004-O Rolled and drawn rod: typical tensile properties

Temperature		Time at temperature, h	Tensile strength		Yield strength		Elongation in 4D, %	Modulus of elasticity(a)	
°F	°C		ksi	MPa	ksi	MPa		10^6 psi	GPa
−320	−196	…	42	290	13	90	38	…	…
−112	−80	…	28	195	11	75	30	…	…
−18	−28	…	26	180	10	70	26	…	…
75	25	…	26	180	10	70	25	10	68
212	100	0.1	26	180	10	70	25	10	66
		0.5	26	180	10	70	25	10	66
		10	26	180	10	70	25	10	66
		100	26	180	10	70	25	10	66
		1,000	26	180	10	70	25	10	66
		10,000	26	180	10	70	25	10	66
300	150	0.1	25	170	10	70	30	9	62
		0.5	25	170	10	70	30	9	62
		10	25	170	10	70	30	9	62
		100	25	170	10	70	30	9	62
		1,000	24	165	10	70	35	9	62
		10,000	22	150	10	70	35	9	62
350	177	0.1	22	150	10	70	50	9	59
		0.5	22	150	10	70	50	9	59
		10	22	150	10	70	50	9	59
		100	21	145	10	70	55	9	59
		1,000	19	130	10	70	55	9	59
		10,000	17	115	10	70	55	9	59
400	205	0.1	17	115	10	70	65	8	57
		0.5	17	115	10	70	65	8	57
		10	17	115	10	70	65	8	57
		100	16	110	10	70	65	8	57
		1,000	15	105	10	70	65	8	57
		10,000	15	105	10	70	65	8	57
450	230	0.1	14	95	10	70	75	8	53
		0.5	14	95	10	70	75	8	53
		10	14	95	10	70	75	8	53
		100	13	90	10	70	75	8	53
		1,000	13	90	10	70	75	8	53
		10,000	13	90	10	70	75	8	53
500	260	0.1	11	75	8.5	59	95	7	49
		0.5	11	75	8.5	59	95	7	49
		10	11	75	8.5	59	95	7	49
		100	11	75	8.5	59	95	7	49

(continued)

(a) The modulus of elasticity in compression is about 2% greater than in tension. Source data are in English units; metric values are converted and rounded.

3004-O Rolled and drawn rod (continued)

Temperature		Time at temperature, h	Tensile strength		Yield strength		Elongation in 4D, %	Modulus of elasticity(a)	
°F	°C		ksi	MPa	ksi	MPa		10^6 psi	GPa
		1,000	11	75	8.5	59	95	7	49
		10,000	11	75	8.5	59	95	7	49
600	315	0.1	8.0	55	6	41	100	6	42
		0.5	8.0	55	6	41	100	6	42
		10	8.0	55	6	41	100	6	42
		100	8.0	55	6	41	100	6	42
		1,000	8.0	55	6	41	100	6	42
700	370	0.1	5.5	38	4	28	110	5	34
		0.5	5.5	38	4	28	110	5	34
		10	5.5	38	4	28	110	5	34
		100	5.5	38	4	28	110	5	34
		1,000	5.5	38	4	28	110	5	34
		10,000	5.5	38	4	28	110	5	34

(a) The modulus of elasticity in compression is about 2% greater than in tension. Source data are in English units; metric values are converted and rounded.

4032–T6: typical tensile properties

Temperature		Time at temperature, h	Tensile strength		Yield strength		Elongation in 4D, %	Modulus of elasticity(a)	
°F	°C		ksi	MPa	ksi	MPa		10^6 psi	GPa
−320	−196	...	66	455	48	330	11	...	...
−112	−80	...	58	400	46	315	10	...	...
−18	−28	...	56	385	46	315	9	...	...
75	25	...	55	380	46	315	9	11.3	78
212	100	0.5	50	345	44	305	9	11.0	76
		10	50	345	44	305	9	11.0	76
		100	50	345	44	305	9	11.0	76
		1,000	51	350	44	305	9	11.0	76
		10,000	50	350	44	310	9	11.0	76
300	150	0.5	46	315	41	285	9	10.5	72
		10	47	325	42	290	9	10.5	72
		100	47	325	42	290	9	10.5	72
		1,000	44	305	40	275	9	10.5	72
		10,000	37	255	33	230	9	10.5	72
350	177	0.5	43	295	38	260	9	10.3	71
		10	43	295	39	270	9	10.3	71
		100	40	275	36	250	10	10.3	71

(continued)

(a) The modulus of elasticity in compression is about 2% greater than in tension. Source data are in English units; metric values are converted and rounded.

4032–T6 (continued)

Temperature		Time at temperature, h	Tensile strength		Yield strength		Elongation in 4D, %	Modulus of elasticity(a)	
°F	°C		ksi	MPa	ksi	MPa		10^6 psi	GPa
		1,000	34	235	27	185	12	10.3	71
		10,000	20	140	13	90	16	10.3	71
400	205	0.5	39	270	35	240	9	10.1	70
		10	36	250	32	220	10	10.1	70
		100	30	205	26	180	15	10.1	70
		1,000	20	140	16	110	22	10.1	70
		10,000	13	90	9.0	62	30	10.1	70
450	230	0.5	32	220	29	200	10	9.7	67
		10	28	195	25	170	13	9.7	67
		100	20	140	17	115	21	9.7	67
		1,000	13	90	10	70	35	9.7	67
		10,000	10	70	7.0	48	40	9.7	67
500	260	0.5	25	170	22	150	12	9.3	64
		10	20	140	18	125	18	9.3	64
		100	14	95	11	75	30	9.3	64
		1,000	10	70	7.0	48	45	9.3	64
		10,000	8.0	55	5.5	38	50	9.3	64
600	315	0.5	13	90	10	70	26	7.8	54
		10	10	70	7.5	52	40	7.8	54
		100	7.5	52	5.0	34	55	7.8	54
		1,000	6.0	41	3.8	26	65	7.8	54
		10,000	5.0	34	3.2	22	70	7.8	54
700	370	0.5	6.5	45	5.0	34	65	6.0	41
		10	4.9	34	3.5	24	80	6.0	41
		100	3.9	27	2.6	18	90	6.0	41
		1,000	3.4	23	2.2	15	90	6.0	41
		10,000	3.4	23	2.0	14	90	6.0	41

(a) The modulus of elasticity in compression is about 2% greater than in tension. Source data are in English units; metric values are converted and rounded.

5050-O: typical tensile properties

Temperature		Time at temperature, h	Tensile strength		Yield strength		Elongation in 4D, %	Modulus of elasticity(a)	
°F	°C		ksi	MPa	ksi	MPa		10^6 psi	GPa
−320	−196	...	37	225	10	70	...	...	...
−112	−80	...	22	150	8.5	60	...	...	...
−18	−28	...	21	145	8.0	55	...	...	...
75	25	...	21	145	8.0	55	...	9.9	68
212	100	0.5	21	145	8.0	55	...	...	...
		10	21	145	8.0	55	...	...	...
		100	21	145	8.0	55	...	...	...
		1,000	21	145	8.0	55	...	...	...
		10,000	21	145	8.0	55	...	...	...
300	150	0.5	19	130	8.0	55	...	...	...
		10	19	130	8.0	55	...	...	...
		100	19	130	8.0	55	...	...	...
		1,000	19	130	8.0	55	...	...	...
		10,000	19	130	8.0	55	...	...	...
350	177	0.5	17	115	8.0	55	...	...	...
		10	17	115	8.0	55	...	...	...
		100	17	115	8.0	55	...	...	...
		1,000	17	115	8.0	55	...	...	...
		10,000	17	115	8.0	55	...	...	...
400	205	0.5	14	95	7.5	52	60	...	...
		10	14	95	7.5	52	60	...	...
		100	14	95	7.5	52	60	...	...
		1,000	14	95	7.5	52	60	...	...
		10,000	14	95	7.5	52	60	...	...
450	230	0.5	11	75	7.0	48	70	...	...
		10	11	75	7.0	48	70	...	...
		100	11	75	7.0	48	70	...	...
		1,000	11	75	7.0	48	70	...	...
		10,000	11	75	7.0	48	70	...	...
500	260	0.5	9.0	62	6.0	41	80	...	...
		10	9.0	62	6.0	41	80	...	...
		100	9.0	62	6.0	41	80	...	...
		1,000	9.0	62	6.0	41	80	...	...
		10,000	9.0	62	6.0	41	80	...	...
600	315	0.5	6.0	41	4.2	29	110	...	...
		10	6.0	41	4.2	29	110	...	...
		100	6.0	41	4.2	29	110	...	...
		1,000	6.0	41	4.2	29	110	...	...

(continued)

(a) The modulus of elasticity in compression is about 2% greater than in tension. (b) Elongation in 2 in. (50 mm) for sheet-type specimens. Source data are in English units; metric values are converted and rounded.

5050-O (continued)

Temperature		Time at temperature, h	Tensile strength		Yield strength		Elongation in 4D, %	Modulus of elasticity(a)	
°F	°C		ksi	MPa	ksi	MPa		10^6 psi	GPa
		10,000	6.0	41	4.2	29	110	...	...
700	370	0.5	3.9	27	2.6	18	130	...	...
		10	3.9	27	2.6	18	130	...	...
		100	3.9	27	2.6	18	130	...	...
		1,000	3.9	27	2.6	18	130	...	...
		10,000	3.9	27	2.6	18	130	...	...

(a) The modulus of elasticity in compression is about 2% greater than in tension. (b) Elongation in 2 in. (50 mm) for sheet-type specimens. Source data are in English units; metric values are converted and rounded.

5052-O: typical tensile properties

Temperature		Time at temperature, h	Tensile strength		Yield strength		Elongation in 4D, %	Modulus of elasticity(a)	
°F	°C		ksi	MPa	ksi	MPa		10^6 psi	GPa
−320	−196	...	44	305	15	105	46	11.3	78
−112	−80	...	29	200	13	90	35	10.6	73
−18	−28	...	28	195	13	90	32	10.3	71
75	25	...	28	195	13	90	30	10.1	70
212	100	0.1	28	195	13	90	36	9.6	66
		0.5	28	195	13	90	36	9.6	66
		10	28	195	13	90	36	9.6	66
		100	28	195	13	90	36	9.6	66
		1,000	28	195	13	90	36	9.6	66
		10,000	28	195	13	90	36	9.6	66
		100,000	28	195	13	90	36	9.6	66
300	150	0.1	23	160	13	90	50	8.9	61
		0.5	23	160	13	90	50	8.9	61
		10	23	160	13	90	50	8.9	61
		100	23	160	13	90	50	8.9	61
		1,000	23	160	13	90	50	8.9	61
		10,000	23	160	13	90	50	8.9	61
		100,000	23	160	13	90	50	8.9	61
350	177	0.1	20	140	12	85	55	8.4	58
		0.5	20	140	12	85	55	8.4	58
		10	20	140	12	85	55	8.4	58
		100	20	140	12	85	55	8.4	58
		1,000	20	140	12	85	55	8.4	58
		10,000	20	140	12	85	55	8.4	58
		100,000	20	140	12	85	55	8.4	58

(continued)

(a) The modulus of elasticity in compression is about 2% greater than in tension. Source data are in English units; metric values are converted and rounded.

5052-O (continued)

Temperature		Time at temperature, h	Tensile strength		Yield strength		Elongation in 4D, %	Modulus of elasticity(a)	
°F	°C		ksi	MPa	ksi	MPa		10^6 psi	GPa
400	205	0.1	17	115	11	75	60	7.9	54
		0.5	17	115	11	75	60	7.9	54
		10	17	115	11	75	60	7.9	54
		100	17	115	11	75	60	7.9	54
		1,000	17	115	11	75	60	7.9	54
		10,000	17	115	11	75	60	7.9	54
		100,000	17	115	11	75	60	7.9	54
450	230	0.1	14	95	9.5	66	70	7.4	51
		0.5	14	95	9.5	66	70	7.4	51
		10	14	95	9.5	66	70	7.4	51
		100	14	95	9.5	66	70	7.4	51
		1,000	14	95	9.5	66	70	7.4	51
		10,000	14	95	9.5	66	70	7.4	51
		100,000	14	95	9.5	66	70	7.4	51
500	260	0.1	12	85	7.5	52	80	6.9	48
		0.5	12	85	7.5	52	80	6.9	48
		10	12	85	7.5	52	80	6.9	48
		100	12	85	7.5	52	80	6.9	48
		1,000	12	85	7.5	52	80	6.9	48
		10,000	12	85	7.5	52	80	6.9	48
		100,000	12	85	7.5	52	80	6.9	48
600	315	0.1	7.5	52	5.5	38	110	5.8	40
		0.5	7.5	52	5.5	38	110	5.8	40
		10	7.5	52	5.5	38	110	5.8	40
		100	7.5	52	5.5	38	110	5.8	40
		1,000	7.5	52	5.5	38	110	5.8	40
		10,000	7.5	52	5.5	38	110	5.8	40
		100,000	7.5	52	5.5	38	110	5.8	40
700	370	0.1	5.0	34	3.1	21	130	4.6	32
		0.5	5.0	34	3.1	21	130	4.6	32
		10	5.0	34	3.1	21	130	4.6	32
		100	5.0	34	3.1	21	130	4.6	32
		1,000	5.0	34	3.1	21	130	4.6	32
		10,000	5.0	34	3.1	21	130	4.6	32
		100,000	5.0	34	3.1	21	130	4.6	32
800	425	...	3.4	23	2.1	14	135	...	...
900	480	...	2.2	15	1.6	11	135	...	...
1000	540	...	1.4	10	1.2	8.0	110	...	...

(a) The modulus of elasticity in compression is about 2% greater than in tension. Source data are in English units; metric values are converted and rounded.

5083-O: typical tensile properties

Temperature		Time at temperature, h	Tensile strength		Yield strength		Elongation in 4D, %	Modulus of elasticity(a)	
°F	°C		ksi	MPa	ksi	MPa		10^6 psi	GPa
−452	−269	...	79	545	27	185	30	...	...
−423	−253	...	77	530	26	180	30	...	...
−320	−196	...	59	405	23	160	34	11.4	79
−112	−80	...	43	295	21	145	27	10.7	74
−18	−28	...	42	290	21	145	24	10.4	72
75	25	...	42	290	21	145	22	10.2	70
212	100	0.1	41	285	21	145	28	9.8	68
		0.5	41	285	21	145	28	9.8	68
		10	41	285	21	145	28	9.8	68
		100	41	285	21	145	28	9.8	68
		1,000	41	285	21	145	29	9.8	68
		10,000	40	275	21	145	34	9.8	68
300	150	0.1	35	240	21	145	45	9.4	65
		0.5	35	240	21	145	45	9.4	65
		10	35	240	21	145	45	9.4	65
		100	35	240	21	145	45	9.4	65
		1,000	33	230	20	140	45	9.4	65
		10,000	31	215	19	130	50	9.4	65
		100,000	...	...	...	...	...	...	...
350	177	0.1	31	215	21	145	50	9.0	62
		0.5	31	215	21	145	50	9.0	62
		10	31	215	21	145	50	9.0	62
		100	30	205	21	145	50	9.0	62
		1,000	28	195	19	130	55	9.0	62
		10,000	26	180	18	125	55	9.0	62
400	205	0.1	...	...	...	...	...	8.6	59
		0.5	27	185	20	140	55	8.6	59
		10	26	180	20	140	55	8.6	59
		100	25	170	20	140	55	8.6	59
		1,000	23	160	18	125	60	8.6	59
		10,000	22	150	17	115	60	8.6	59
450	230	0.1	...	...	...	...	...	8.1	56
		0.5	21	145	15	105	60	8.1	56
		10	21	145	14	95	60	8.1	56
		100	20	140	14	95	65	8.1	56
		1,000	20	140	14	95	65	8.1	56
		10,000	20	140	14	95	65	8.1	56
		100,000	20	140	14	95	65	8.1	56

(continued)

(a) The modulus of elasticity in compression is about 2% greater than in tension. Source data are in English units; metric values are converted and rounded.

5083-O (continued)

Temperature		Time at temperature, h	Tensile strength		Yield strength		Elongation in 4D, %	Modulus of elasticity(a)	
°F	°C		ksi	MPa	ksi	MPa		10^6 psi	GPa
500	260	0.1	...	...	...	...	...	7.6	52
		0.5	17	115	11	75	80	7.6	52
		10	17	115	11	75	80	7.6	52
		100	17	115	11	75	80	7.6	52
		1,000	17	115	11	75	80	7.6	52
		10,000	17	115	11	75	80	7.6	52
		100,000	17	115	11	75	80	7.6	52
600	315	0.1	11	75	7.5	52	110	6.4	44
		0.5	11	75	7.5	52	110	6.4	44
		10	11	75	7.5	52	110	6.4	44
		100	11	75	7.5	52	110	6.4	44
		1,000	11	75	7.5	52	110	6.4	44
		10,000	11	75	7.5	52	110	6.4	44
		100,000	11	75	7.5	52	110	6.4	44
700	370	0.1	6.0	41	4.2	29	130	5.2	36
		0.5	6.0	41	4.2	29	130	5.2	36
		10	6.0	41	4.2	29	130	5.2	36
		100	6.0	41	4.2	29	130	5.2	36
		1,000	6.0	41	4.2	29	130	5.2	36
		10,000	6.0	41	4.2	29	130	5.2	36
		100,000	6.0	41	4.2	29	130	5.2	36
800	425	...	3.4	23	2.1	14	135	...	...
900	480	...	2.2	15	1.6	11	135	...	...
1000	540	...	1.4	10	1.2	8.0	110	...	...

(a) The modulus of elasticity in compression is about 2% greater than in tension. Source data are in English units; metric values are converted and rounded.

5086-O: typical tensile properties

Temperature		Time at temperature, h	Tensile strength		Yield strength		Elongation in 4D, %	Modulus of elasticity(a)	
°F	°C		ksi	MPa	ksi	MPa		10^6 psi	GPa
−423	−253	...	79	545	22	150	42	...	...
−320	−196	...	55	380	19	130	46	11.5	79
−112	−80	...	39	270	17	115	35	10.8	74
−18	−28	...	38	260	17	115	32	10.5	72
75	25	...	38	260	17	115	30	10.3	71
212	100	0.5	38	260	17	115	36	9.9	68
		10	38	260	17	115	36	9.9	68

(continued)

(a) Average of tensile and compressive moduli. Source data are in English units; metric values are converted and rounded.

5086-O (continued)

Temperature		Time at temperature, h	Tensile strength		Yield strength		Elongation in 4D, %	Modulus of elasticity(a)	
°F	°C		ksi	MPa	ksi	MPa		10^6 psi	GPa
		100	38	260	17	115	36	9.9	68
		1,000	38	260	17	115	36	9.9	68
		10,000	38	260	17	115	36	9.9	68
300	150	0.5	32	220	16	110	50	9.5	66
		10	32	220	16	110	50	9.5	66
		100	32	220	16	110	50	9.5	66
		1,000	30	205	16	110	50	9.5	66
		10,000	29	200	16	110	50	9.5	66
350	177	0.5	27	185	16	110	55	9.1	63
		10	26	180	16	110	55	9.1	63
		100	26	180	16	110	55	9.1	63
		1,000	26	180	16	110	55	9.1	63
		10,000	26	180	16	110	55	9.1	63
400	205	0.5	22	150	15	105	60	8.7	60
		10	22	150	15	105	60	8.7	60
		100	22	150	15	105	60	8.7	60
		1,000	22	150	15	105	60	8.7	60
		10,000	22	150	15	105	60	8.7	60
500	260	0.5	17	115	11	75	80	7.7	53
		10	17	115	11	75	80	7.7	53
		100	17	115	11	75	80	7.7	53
		1,000	17	115	11	75	80	7.7	53
		10,000	17	115	11	75	80	7.7	53
600	315	0.5	11	75	7.5	52	110	6.5	45
		10	11	75	7.5	52	110	6.5	45
		100	11	75	7.5	52	110	6.5	45
		1,000	11	75	7.5	52	110	6.5	45
		10,000	11	75	7.5	52	110	6.5	45
700	370	0.5	6.0	41	4.2	29	130	5.2	36
		10	6.0	41	4.2	29	130	5.2	36
		100	6.0	41	4.2	29	130	5.2	36
		1,000	6.0	41	4.2	29	130	5.2	36
		10,000	6.0	41	4.2	29	130	5.2	36

(a) Average of tensile and compressive moduli. Source data are in English units; metric values are converted and rounded.

5454-O: typical tensile properties

Temperature		Time at temperature, h	Tensile strength		Yield strength		Elongation in 4D, %	Modulus of elasticity(a)	
°F	°C		ksi	MPa	ksi	MPa		10^6 psi	GPa
−452	−269	...	74	510	24	165	34	...	...
−320	−196	...	54	370	19	140	38	...	...
−112	−80	...	37	255	17	115	31	...	...
−18	−28	...	36	250	17	115	28	...	...
75	25	...	36	250	17	115	25	10.1	70
212	100	0.5	36	250	17	115	31	9.6	66
		10	36	250	17	115	31	9.6	66
		100	36	250	17	115	31	9.6	66
		1,000	36	250	17	115	31	9.6	66
		10,000	36	250	17	115	31	9.6	66
300	150	0.5	29	200	16	110	50	8.9	61
		10	29	200	16	110	50	8.9	61
		100	29	200	16	110	50	8.9	61
		1,000	29	200	16	110	50	8.9	61
		10,000	29	200	16	110	50	8.9	61
350	177	0.5	26	180	16	110	55	8.4	58
		10	26	180	16	110	55	8.4	58
		100	26	180	16	110	55	8.4	58
		1,000	26	180	16	110	55	8.4	58
		10,000	26	180	16	110	55	8.4	58
400	205	0.5	22	150	15	105	60	7.9	54
		10	22	150	15	105	60	7.9	54
		100	22	150	15	105	60	7.9	54
		1,000	22	150	15	105	60	7.9	54
		10,000	22	150	15	105	60	7.9	54
450	230	0.5	20	140	13	90	70	7.4	51
		10	20	140	13	90	70	7.4	51
		100	20	140	13	90	70	7.4	51
		1,000	20	140	13	90	70	7.4	51
		10,000	20	140	13	90	70	7.4	51
500	260	0.5	17	115	11	75	80	6.9	48
		10	17	115	11	75	80	6.9	48
		100	17	115	11	75	80	6.9	48
		1,000	17	115	11	75	80	6.9	48
		10,000	17	115	11	75	80	6.9	48
600	315	0.5	11	75	7.5	52	110	5.8	40
		10	11	75	7.5	52	110	5.8	40
		100	11	75	7.5	52	110	5.8	40
		1,000	11	75	7.5	52	110	5.8	40

(continued)

(a) The modulus of elasticity in compression is about 2% greater than in tension. Source data are in English units; metric values are converted and rounded.

5454-O (continued)

Temperature		Time at temperature, h	Tensile strength		Yield strength		Elongation in 4D, %	Modulus of elasticity(a)	
°F	°C		ksi	MPa	ksi	MPa		10^6 psi	GPa
		10,000	11	75	7.5	52	110	5.8	40
700	370	0.5	6.0	41	4.2	29	130	4.6	32
		10	6.0	41	4.2	29	130	4.6	32
		100	6.0	41	4.2	29	130	4.6	32
		1,000	6.0	41	4.2	29	130	4.6	32
		10,000	6.0	41	4.2	29	130	4.6	32

(a) The modulus of elasticity in compression is about 2% greater than in tension. Source data are in English units; metric values are converted and rounded.

6061-T6, -T651, -T6511 (except for T6 sheet and rolled-and-drawn products): typical tensile properties

Temperature		Time at temperature, h	Tensile strength		Yield strength		Elongation in 4D, %	Modulus of elasticity(a)	
°F	°C		ksi	MPa	ksi	MPa		10^6 psi	GPa
−452	−269	...	74	510	53	365	27	...	...
−423	−253	...	74	510	53	365	27	...	...
−320	−196	...	60	415	47	325	22	11.1	77
−112	−80	...	49	340	42	290	19	10.4	72
−18	−28	...	47	325	41	285	18	10.1	70
75	25	...	45	310	40	275	17	9.9	68
212	100	0.1	41	285	38	260	18	9.5	66
		0.5	41	285	38	260	18	9.5	66
		10	41	285	38	260	18	9.5	66
		100	41	285	38	260	18	9.5	66
		1,000	42	290	39	270	18	9.5	66
		10,000	42	290	39	270	18	9.5	66
		100,000	42	290	39	270	17	9.5	66
300	150	0.1	38	260	36	250	20	9.1	63
		0.5	38	260	36	250	20	9.1	63
		10	38	260	36	250	19	9.1	63
		100	38	260	36	250	18	9.1	63
		1,000	38	260	36	250	18	9.1	63
		10,000	33	230	31	215	19	9.1	63
		100,000	29	200	27	185	21	9.1	63
350	177	0.1	36	250	35	240	21	8.9	61

(continued)

(a) The modulus of elasticity in compression is about 2% greater than in tension. Source data are in English units; metric values are converted and rounded.

6061-T6, -T651, -T6511 (except for T6 sheet and rolled-and-drawn products) (continued)

Temperature		Time at temperature, h	Tensile strength		Yield strength		Elongation in 4D, %	Modulus of elasticity(a)	
°F	°C		ksi	MPa	ksi	MPa		10^6 psi	GPa
		0.5	36	250	35	240	21	8.9	61
		10	36	250	35	240	18	8.9	61
		100	34	235	32	220	18	8.9	61
		1,000	30	205	28	195	19	8.9	61
		10,000	24	165	22	150	22	8.9	61
		100,000	20	140	17	115	28	8.9	61
400	205	0.1	33	230	32	220	20	8.6	59
		0.5	34	235	32	220	19	8.6	59
		10	30	205	30	205	18	8.6	59
		100	27	185	25	170	19	8.6	59
		1,000	22	150	20	140	23	8.6	59
		10,000	17	115	14	95	28	8.6	59
		100,000	13	90	9.0	62	40	8.6	59
450	230	0.1	29	200	28	195	19	8.3	57
		0.5	29	200	28	195	17	8.3	57
		10	24	165	23	160	18	8.3	57
		100	19	130	18	125	23	8.3	57
		1,000	15	105	13	90	30	8.3	57
		10,000	12	85	8.5	59	40	8.3	57
		100,000	9.0	62	6.5	45	70	8.3	57
500	260	0.1	25	170	24	165	17	7.9	54
		0.5	23	160	22	150	16	7.9	54
		10	18	125	16	110	20	7.9	54
		100	13	90	11	75	29	7.9	54
		1,000	10	70	8.0	55	45	7.9	54
		10,000	8.0	55	6.0	41	65	7.9	54
		100,000	7.0	48	5.5	38	80	7.9	54
600	315	0.1	14	95	13.0	90	18	6.8	47
		0.5	12	85	11.0	75	23	6.8	47
		10	9.0	62	8.0	55	30	6.8	47
		100	6.0	41	4.5	31	65	6.8	47
		1,000	5.0	34	4.2	29	80	6.8	47
		10,000	5.0	34	4.2	29	80	6.8	47
		100,000	5.0	34	4.2	29	80	6.8	47
700	370	0.1	8.5	59	8.0	55	35	5.5	38
		0.5	7.0	48	6.5	45	35	5.5	38
		10	3.8	26	3.0	21	80	5.5	38

(continued)

(a) The modulus of elasticity in compression is about 2% greater than in tension. Source data are in English units; metric values are converted and rounded.

6061-T6, -T651, -T6511 (continued)

Temperature		Time at temperature, h	Tensile strength		Yield strength		Elongation in 4D, %	Modulus of elasticity(a)	
°F	°C		ksi	MPa	ksi	MPa		10^6 psi	GPa
		100	3.6	25	3.0	21	80	5.5	38
		1,000	3.6	25	3.0	21	80	5.5	38
		10,000	3.6	25	3.0	21	80	5.5	38
		100,000	3.6	25	3.0	21	80	5.5	38
800	425	0.1	3.8	26	3.2	22	65	...	...
		0.5	3.0	21	2.4	17	80	...	...
900	480	...	2.2	15	1.6	11	80	...	...
1000	540	...	1.6	11	1.2	8.0	65	...	...

(a) The modulus of elasticity in compression is about 2% greater than in tension. Source data are in English units; metric values are converted and rounded.

6063-T6: typical tensile properties

Temperature		Time at temperature, h	Tensile strength		Yield strength		Elongation in 4D, %	Modulus of elasticity(a)	
°F	°C		ksi	MPa	ksi	MPa		10^6 psi	GPa
−320	−196	...	47	325	36	250	24	...	...
−112	−80	...	38	260	33	230	20	...	...
−18	−28	...	36	250	32	220	19	...	...
75	25	...	35	240	31	215	18	9.9	68
212	100	0.5	31	215	28	195	20	...	...
		10	31	215	28	195	20	...	...
		100	31	215	29	200	20	...	...
		1,000	32	220	30	205	19	...	...
		10,000	33	230	31	215	15	...	...
300	150	0.5	28	195	26	180	22	...	...
		10	28	195	26	180	20	...	...
		100	29	200	28	195	18	...	...
		1,000	28	195	27	185	15	...	...
		10,000	22	150	20	140	20	...	...
350	177	0.5	26	180	25	170	19	...	...
		10	25	170	24	165	17	...	...
		100	24	165	23	160	16	...	...
		1,000	20	140	18	125	16	...	...
		10,000	14	95	11	75	29	...	...
400	205	0.5	24	165	23	160	18	...	...

(continued)

(a) The modulus of elasticity in compression is about 2% greater than in tension. Source data are in English units; metric values are converted and rounded.

6063-T6 (conitnued)

Temperature		Time at temperature, h	Tensile strength		Yield strength		Elongation in 4D, %	Modulus of elasticity(a)	
°F	°C		ksi	MPa	ksi	MPa		10^6 psi	GPa
		10	20	140	18	125	18	...	...
		100	16	110	15	105	24	...	...
		1,000	13	90	10	70	30	...	...
		10,000	9.0	62	6.5	45	40	...	...
450	230	0.5	20	140	18	125	20	...	...
		10	15	105	13	90	22	...	...
		100	11	75	9.5	66	35	...	...
		1,000	8.0	55	6.0	41	50	...	...
		10,000	6.0	41	4.5	31	60	...	...
500	260	0.5	14	95	12	85	25	...	...
		10	10	70	9.0	62	30	...	...
		100	7.5	52	6.0	41	45	...	...
		1,000	5.0	34	4.0	28	65	...	...
		10,000	4.5	31	3.5	24	75	...	...
600	315	0.5	5.5	38	5.0	34	38	...	...
		10	4.0	28	3.5	24	55	...	...
		100	3.5	24	3.0	21	70	...	...
		1,000	3.2	22	2.5	17	75	...	...
		10,000	3.2	22	2.5	17	80	...	...
700	370	0.5	2.3	16	2.0	14	90	...	...
		10	2.3	16	2.0	14	100	...	...
		100	2.3	16	2.0	14	105	...	...
		1,000	2.3	16	2.0	14	105	...	...
		10,000	2.3	16	2.0	14	105	...	...

(a) The modulus of elasticity in compression is about 2% greater than in tension. Source data are in English units; metric values are converted and rounded.

6101-T6: typical tensile properties

Temperature		Time at temperature, h	Tensile strength		Yield strength		Elongation in 4D, %	Modulus of elasticity(a)	
°F	°C		ksi	MPa	ksi	MPa		10^6 psi	GPa
−320	−196	...	43	295	33	230	24	...	...
−112	−80	...	36	250	30	205	20	...	...
−18	−28	...	34	235	29	200	19	...	...
75	25	...	32	220	28	195	19	9.9	68
212	100	0.5	28	195	25	170	20	...	...
		10	28	195	25	170	20	...	...
		100	28	195	25	170	20	...	...

(continued)

(a) The modulus of elasticity in compression is about 2% greater than in tension. Source data are in English units; metric values are converted and rounded.

6101-T6 (continued)

Temperature		Time at temperature, h	Tensile strength		Yield strength		Elongation in 4D, %	Modulus of elasticity(a)	
°F	°C		ksi	MPa	ksi	MPa		10^6 psi	GPa
		1,000	28	195	25	170	20	...	...
		10,000	28	195	25	170	20	...	...
300	150	0.5	25	170	23	160	20	...	...
		10	25	170	23	160	20	...	...
		100	25	170	23	160	20	...	...
		1,000	24	165	22	150	20	...	...
		10,000	21	145	19	130	20	...	...
350	177	0.5	24	165	22	150	20	...	...
		10	23	160	22	150	20	...	...
		100	21	145	20	140	21	...	...
		1,000	19	130	16	110	24	...	...
		10,000	15	105	12	85	30	...	...
400	205	0.5	21	145	20	140	21	...	...
		10	19	130	18	125	21	...	...
		100	16	110	14	95	24	...	...
		1,000	13	90	11	75	30	...	...
		10,000	10	70	7.0	48	40	...	...
450	230	0.5	18	125	16	110	24	...	...
		10	15	105	13	90	27	...	...
		100	12	85	9.5	66	35	...	...
		1,000	8.5	59	6.5	45	45	...	...
		10,000	7.0	48	4.8	33	55	...	...
500	260	0.5	14	95	12	85	29	...	...
		10	11	75	8.5	59	35	...	...
		100	8.0	55	6.0	41	50	...	...
		1,000	5.5	38	4.1	28	70	...	...
		10,000	4.8	33	3.3	23	80	...	...
600	315	0.5	7.0	48	6.0	41	50	...	...
		10	4.7	32	3.6	25	70	...	...
		100	3.8	26	2.9	20	85	...	...
		1,000	3.3	23	2.6	18	90	...	...
		10,000	3.0	21	2.3	16	100	...	...
700	370	0.5	2.5	17	1.8	12	85	...	...
		10	2.5	17	1.8	12	105	...	...
		100	2.5	17	1.8	12	105	...	...
		1,000	2.5	17	1.8	12	105	...	...
		10,000	2.5	17	1.8	12	105	...	...

(a) The modulus of elasticity in compression is about 2% greater than in tension. Source data are in English units; metric values are converted and rounded.

7005-T53 Extrusions: typical tensile properties

Temperature		Time at temperature, h	Tensile strength		Yield strength		Elongation in 4D, %	Modulus of elasticity(a)	
°F	°C		ksi	MPa	ksi	MPa		10^6 psi	GPa
−452	−269	...	93	641	70	485	16	...	...
320	−196	...	78	540	61	420	16	...	...
112	−80	...	64	440	55	380	13	...	...
18	−28	...	61	420	52	360	14	...	...
75	25	...	57	395	50	345	15	10.3	71
212	100	0.1	49	340	46	315	19	...	...
		0.5	49	340	46	315	19	...	...
		10	49	340	46	315	19	...	...
		100	50	345	47	325	18	...	...
		1,000	50	345	48	330	18	...	...
		10,000	44	305	41	285	20	...	...
300	150	0.1	43	295	41	285	23	...	...
		0.5	44	305	42	290	24	...	...
		10	43	295	42	290	23	...	...
		100	37	255	36	250	24	...	...
		1,000	30	205	29	200	25	...	...
		10,000	24	165	21	145	35	...	...
350	177	0.1	39	270	37	255	26	...	...
		0.5	38	260	37	255	25	...	...
		10	32	220	31	215	27	...	...
		100	26	180	25	170	30	...	...
		1,000	20	140	18	125	40	...	...
		10,000	16	110	14	95	50	...	...
400	205	0.1	32	220	31	215	30	...	...
		0.5	30	205	29	200	30	...	...
		10	23	160	22	150	35	...	...
		100	18	125	16	110	40	...	...
		1,000	15	105	13	90	55	...	...
		10,000	14	95	12	85	60	...	...
450	230	0.1	24	165	23	160	35	...	...
		0.5	22	150	21	145	35	...	...
		10	16	110	15	105	45	...	...
		100	14	95	12	85	55	...	...
		1,000	13	90	11	75	70	...	...
		10,000	12	85	10	70	70	...	...
500	260	0.1	17	115	16	110	45	...	...
		0.5	15	105	14	95	45	...	...
		10	12	85	11	75	55	...	...
		100	11	75	10	70	65	...	...

(continued)

(a) The modulus of elasticity in compression is about 2% greater than in tension. (b) Cooled in still air at room temperature and treated within 2 h after removal from holding oven. This property may increase with time after cooling to room temperature. Source data are in English units; metric values are converted and rounded.

7005-T53 Extrusions (continued)

Temperature		Time at temperature, h	Tensile strength		Yield strength		Elongation in 4D, %	Modulus of elasticity(a)	
°F	°C		ksi	MPa	ksi	MPa		10^6 psi	GPa
		1,000	11	75	10	70	80	...	...
		10,000	11	75	9.5	66	80	...	...
600	315	0.1	...	...	...	...	...	...	...
		0.5	9.0	62	7.0	48	95	...	...
		10	8.5	59	7.0	48	95	...	...
		100	8.5	59	7.0	48	95	...	...
		1,000	8.5	59	7.0	48	95	...	...
		10,000	8.5	59	7.0	48	95	...	...
700	370	0.1	...	...	...	...	...	...	...
		0.5	6.0	41	4.5	31	105	...	...
		10	6.0	41	4.5	31	105	...	...
		100	6.0	41	4.5	31	105	...	...
		1,000	6.0	41	4.5	31	105	...	...
		10,000	6.0	41	4.5	31	105	...	...
800	425	...	3.8	26	2.6	18	115	...	...
900	480	...	2.6	18	1.8	12	120	...	...
1000	540	...	1.6	11	1.0	7.0	120	...	...

(a) The modulus of elasticity in compression is about 2% greater than in tension. (b) Cooled in still air at room temperature and treated within 2 h after removal from holding oven. This property may increase with time after cooling to room temperature. Source data are in English units; metric values are converted and rounded.

7050-T7451 Plate 1.001 to 2.000 in. (>25.00 ≤ 50.00 mm) thick: typical tensile properties

Temperature		Time at temperature, h	Tensile strength		Yield strength		Elongation in 4D, %
°F	°C		ksi	MPa	ksi	MPa	
75	25	0.1	74	510	66	455	11
212	100	0.1	64	440	62	425	13
		0.5	64	440	62	425	13
		10	64	440	62	425	13
		100	65	450	63	435	13
		1,000	64	440	62	425	14
		10,000	64	440	61	420	15
300	150	0.1	57	395	56	385	16
		0.5	57	395	56	385	17
		10	57	395	56	385	18
		100	52	360	51	350	19
		1,000	42	290	40	275	21
		10,000	32	220	28	195	29
350	177	0.1	52	360	50	345	19

(continued)

Source data are in English units; metric values are converted and rounded.

7050-T7451 Plate 1.001 to 2.000 in. (>25.00 ≤ 50.00 mm) thick (continued)

Temperature		Time at temperature, h	Tensile strength		Yield strength		Elongation in 4D, %
°F	°C		ksi	MPa	ksi	MPa	
		0.5	51	350	50	345	20
		10	47	325	45	310	22
		100	36	250	34	235	25
		1,000	28	195	25	170	31
		10,000	23	160	18	125	40
400	205	0.1	44	305	42	290	22
		0.5	42	290	40	275	23
		10	32	220	30	205	27
		100	24	165	22	150	32
		1,000	19	130	16	110	45
		10,000	17	115	13	90	54

Source data are in English units; metric values are converted and rounded.

7075-T6, -T651 except die forgings >2 in. (>50.00 mm) thick and extrusions: typical tensile properties

Temperature		Time at temperature, h	Tensile strength		Yield strength		Elongation in 4D, %	Modulus of elasticity(a)	
°F	°C		ksi	MPa	ksi	MPa		10^6 psi	GPa
−452	−269	...	115	795	105	725	8	...	...
−423	−253	...	114	785	99	685	8	11.9	82
−320	−196	...	102	705	92	635	8	11.5	79
−112	−80	...	90	620	79	545	11	10.8	74
−18	−28	...	86	595	75	515	11	10.5	72
75	25	...	83	570	73	505	11	10.3	71
212	100	0.5	75	515	69	475	15	9.8	68
		10	76	525	70	485	14	9.8	68
		100	77	530	71	490	14	9.8	68
		1,000	76	525	70	485	14	9.8	68
		10,000	70	485	65	450	14	9.8	68
300	150	0.5	64	440	60	415	21	9.1	63
		10	64	440	60	415	19	9.1	63
		100	56	385	53	365	20	9.1	63
		1,000	41	285	38	260	23	9.1	63
		10,000	31	215	27	185	30	9.1	63
350	177	0.5	54	370	50	345	20	8.7	60
		10	46	315	43	295	23	8.7	60
		100	35	240	32	220	26	8.7	60
		1,000	24	165	22	150	35	8.7	60
		10,000	20	140	18	125	45	8.7	60

(continued)

(a) The modulus of elasticity in compression is about 2% greater than in tension. Source data are in English units; metric values are converted and rounded.

7075-T6, -T651 (continued)

Temperature		Time at temperature, h	Tensile strength		Yield strength		Elongation in 4D, %	Modulus of elasticity(a)	
°F	°C		ksi	MPa	ksi	MPa		10^6 psi	GPa
400	205	0.5	42	290	40	275	18	8.2	57
		10	30	205	28	195	27	8.2	57
		100	22	150	20	140	35	8.2	57
		1,000	18	125	16	110	45	8.2	57
		10,000	16	110	13	90	55	8.2	57
450	230	0.5	30	205	28	195	25	7.7	53
		10	19	130	18	125	35	7.7	53
		100	16	110	14	95	40	7.7	53
		1,000	14	95	12	85	50	7.7	53
		10,000	13	90	11	75	60	7.7	53
500	260	0.5	19	130	18	125	35	7.0	48
		10	14	95	13	90	45	7.0	48
		100	12	85	11	75	50	7.0	48
		1,000	12	85	10	70	55	7.0	48
		10,000	11	75	9.0	62	65	7.0	48
		0.5	10	70	8.0	55	60	5.6	39
		10	9.0	62	7.5	52	65	5.6	39
		100	8.5	59	7.0	48	70	5.6	39
		1,000	8.0	55	6.5	45	70	5.6	39
		10,000	8.0	55	6.5	45	70	5.6	39
700	370	0.5	6.0	41	4.6	32	70	4.0	28
		10	6.0	41	4.6	32	70	4.0	28
		100	6.0	41	4.6	32	70	4.0	28
		1,000	6.0	41	4.6	32	70	4.0	28
		10,000	6.0	41	4.6	32	70	4.0	28
800	425	...	4.4	30	3.4	23	80	...	...
900	480	...	3.0	21	2.2	15	75	...	...
1000	540	...	0.8	6.0	0.3	2.0	1	...	...

(a) The modulus of elasticity in compression is about 2% greater than in tension. Source data are in English units; metric values are converted and rounded.

201.0-17 Sand castings: typical tensile properties

Temperature		Time at temperature, h	Tensile strength		Yield strength		Elongation in 4D, %	Modulus of elasticity(a)	
°F	°C		ksi	MPa	ksi	MPa		10^6 psi	GPa
−452	−269	...	93	640	81	560	7	...	...
−423	−253	...	93	640	79	545	8	...	...
−320	−196	...	89	615	67	460	8	...	...
−112	−80	...	77	530	70	485	6	...	...
−18	−28	...	74	510	87	600	6	...	...
75	25	...	72	495	65	450	6	10.3	71
300	150	100	64	440	57	395	9	...	...
		1,000	60	415	54	370	10	...	...
		10,000	58	400	52	360	6	...	...
350	177	100	54	370	49	340	10	...	...
		1,000	51	350	46	315	8	...	...
		10,000	43	295	37	255	9	...	...
400	205	100	48	330	42	290	10	...	...
		1,000	39	270	33	230	16	...	...
		10,000	24	165	18	125	25	...	...
450	230	1,000	22	150	15	105	25	...	...
		10,000	19	130	13	90	25	...	...
500	260	1,000	16	110	13	90	25	...	...
		10,000	14	95	10	70	32	...	...
600	315	1,000	9.0	62	8.0	55	48	...	...
		10,000	8.0	55	6.0	41	51	...	...

(a) The modulus of elasticity in compression is about 2% greater than in tension. Source data are in English units; metric values are converted and rounded.

249.0-T7: Creep-rupture and creep properties

Temperature		Time at temperature, h	Tensile strength		Yield strength		Elongation in 4D, %	Modulus of elasticity(a)	
°F	°C		ksi	MPa	ksi	MPa		10^6 psi	GPa
452	−269	...	94	650	76	525	10	...	...
−423	−253	...	91	627	75	515	9	...	...
−320	−196	...	81	560	69	475	7	...	...
−112	−80	...	70	485	62	425	6	...	...
−18	−28	...	69	475	60	415	6	...	...
75	25	...	68	470	59	405	6	10.3	71
300	150	0.5	55	380	52	360	10	...	...
		10	57	395	52	360	10	...	...
		100	55	380	50	345	11	...	...
		1,000	47	325	41	285	14	...	...
		10,000	42	290	35	240	14	...	...
350	177	0.5	51	350	48	330	10	...	...
		10	50	345	46	315	11	...	...
		100	45	310	41	285	13	...	...
		1,000	37	255	32	220	14	...	...
		10,000	34	235	28	195	17	...	...
400	205	0.5	46	315	41	285	10	...	...
		10	40	275	36	250	12	...	...
		100	36	250	32	220	15	...	...
		1,000	29	200	25	170	16	...	...
		10,000	27	185	23	160	20	...	...

(a) The modulus of elasticity in compression is about 2% greater than in tension. Source data are in English units; metric values are converted and rounded.

295.0-T6 Sand castings: typical tensile properties

Temperature		Time at temperature, h	Tensile strength		Yield strength		Elongation in 4D, %	Modulus of elasticity(a)	
°F	°C		ksi	MPa	ksi	MPa		10^6 psi	GPa
75	25	...	36	250	24	165	5	10	69
212	100	10,000	34	235	23	160	5	...	...
300	150	10,000	28	195	20	140	5	...	...
400	205	10,000	15	105	9.0	62	15	...	...
500	260	10,000	9.0	62	6.0	41	25	...	...
600	315	10,000	4.0	28	3.0	21	75	...	...
700	370	10,000	2.5	17	1.5	10	100	...	...

(a) Average of tensile and compressive moduli. Source data are in English units; metric values are converted and rounded

354.0-T6, -T61 Permanent mold: typical tensile properties

Temperature		Time at temperature, h	Tensile strength		Yield strength		Elongation in 4D, %	Modulus of elasticity(a)	
°F	°C		ksi	MPa	ksi	MPa		10^6 psi	GPa
−320	−196	...	68	470	49	340	5	...	...
−112	−80	...	58	400	42	290	5	...	...
−18	−28	...	58	400	42	290	5	...	...
75	25	...	55	380	41	285	6	10.6	73
212	100	0.5	50	345	41	285	6	...	...
		10	51	350	41	285	6	...	...
		100	52	360	42	290	6	...	...
		1,000	54	370	45	310	6	...	...
		10,000	60	415	49	340	6	...	...
300	150	0.5	47	325	40	275	6	...	...
		10	50	345	43	295	6	...	...
		100	51	350	46	315	6	...	...
		1,000	49	340	44	305	6	...	...
		10,000	42	290	35	240	6	...	...
350	177	0.5	45	310	39	270	6	...	...
		10	47	325	42	290	6	...	...
		100	43	295	38	260	8	...	...
		1,000	33	230	28	195	13	...	...
		10,000	19	130	14	95	24	...	...
400	205	0.5	42	290	39	270	6	...	...
		10	39	270	36	250	9	...	...
		100	30	205	26	180	17	...	...
		1,000	19	130	15	105	30	...	...
		10,000	15	102	11	75	45	...	...
450	230	0.5	37	255	35	240	9	...	...
		10	28	195	25	170	15	...	...
		100	18	125	14	95	25	...	...
		1,000	14	95	11	75	40	...	...
		10,000	12	85	8.5	59	55	...	...
500	260	0.5	28	195	25	170	16	...	...
		10	17	115	15	105	22	...	...
		100	12	85	9.5	66	35	...	...
		1,000	9.5	66	7.5	52	50	...	...
		10,000	8.5	59	6.0	41	65	...	...
600	315	0.5	13	90	12	85	29	...	...
		10	8.5	59	7.0	48	60	...	...
		100	6.0	41	5.0	34	85	...	...
		1,000	...	...	...	...	...	...	...
		10,000	...	...	...	...	...	...	...

(a) The modulus of elasticity in compression is about 2% greater than in tension. Source data are in English units; metric values are converted and rounded.

355.0-T6 Sand castings: typical tensile properties

Temperature			Tensile strength		Yield strength		
°F	°C	Time at temperature, h	ksi	MPa	ksi	MPa	Elongation in 4D, %
75	25	...	35	240	25	170	3
212	100	0.5	35	240	25	170	3
		10	35	240	25	170	3
		100	36	250	26	180	3
		1,000	37	255	30	205	3
		10,000	40	275	35	240	3
300	150	0.5	34	235	25	170	3
		10	37	255	30	205	3
		100	39	270	34	235	3
		1,000	39	270	36	250	3
		10,000	33	230	29	200	3
350	177	0.5	34	235	25	170	3
		10	38	260	33	230	3
		100	37	255	34	235	3
		1,000	30	205	28	195	3
		10,000	23	160	18	125	5
400	205	0.5	33	230	25	170	3
		10	35	240	30	205	3
		100	30	205	26	180	3
		1,000	20	140	17	115	5
		10,000	15	105	11	75	6
500	260	0.5	22	150	19	130	5
		10	17	115	15	105	6
		100	14	95	11	75	8
		1,000	11	75	9.0	62	13
		10,000	9.5	66	6.0	41	16
600	315	0.5	10	70	9.0	62	10
		10	8.5	59	7.0	48	15
		100	7.5	52	6.0	41	22
		1,000	6.5	45	4.5	31	30
		10,000	6.0	41	3.0	21	36
700	370	0.5	5.5	38	4.5	31	25
		10	5.0	34	3.5	24	30
		100	4.5	31	3.0	21	40
		1,000	4.0	28	2.5	17	45
		10,000	3.5	24	2.0	14	50

Source data are in English units; metric values are converted and rounded.

356.0-16 Permanent mold: typical tensile properties

Temperature		Time at temperature, h	Tensile strength		Yield strength		Elongation in 4D, %	Modulus of elasticity(a)	
°F	°C		ksi	MPa	ksi	MPa		10^6 psi	GPa
−423	−253	...	61	420	36	250	3	...	...
−320	−196	...	47	325	32	220	3	...	...
−112	−80	...	43	295	30	205	4	...	...
−18	−28	...	41	285	28	195	4	...	...
75	25	...	38	260	27	185	5	10.4	72
212	100	0.5	34	235	27	185	6	10.2	70
		10	36	250	27	185	6	10.2	70
		100	36	250	27	185	6	10.2	70
		1,000	33	230	26	180	6	10.2	70
		10,000	30	205	25	170	6	10.2	70
300	150	0.5	29	200	25	170	7	9.9	68
		10	32	220	26	180	7	9.9	68
		100	29	200	25	170	8	9.9	68
		1,000	23	160	21	145	9	9.9	68
		10,000	21	145	17	115	10	9.9	68
350	177	0.5	26	180	23	160	8	9.7	67
		10	27	185	24	165	10	9.7	67
		100	23	160	20	140	13	9.7	67
		1,000	18	125	15	105	15	9.7	67
		10,000	16	110	12	85	17	9.7	67
400	205	0.5	23	160	20	140	9	9.4	65
		10	21	145	19	130	15	9.4	65
		100	16	110	13	90	20	9.4	65
		1,000	13	90	9.5	66	25	9.4	65
		10,000	12	85	8.5	59	30	9.4	65
450	230	0.5	19	130	17	115	11	8.9	61
		10	16	110	13	90	21	8.9	61
		100	12	85	9.0	62	29	8.9	61
		1,000	10	70	7.5	52	35	8.9	61
		10,000	9.5	66	6.5	45	45	8.9	61
500	260	0.5	15	105	13	90	17	8.3	57
		10	11	75	9.0	62	28	8.3	57
		100	9.0	62	7.0	48	38	8.3	57
		1,000	8.0	55	6.0	41	46	8.3	57
		10,000	7.5	52	5.0	34	55	8.3	57
600	315	0.5	7.5	52	6.5	45	40	6.8	47
		10	6.5	45	5.0	34	45	6.8	47
		100	5.5	38	4.0	28	50	6.8	47
		1,000	5.0	34	3.5	24	60	6.8	47

(continued)

(a) The modulus of elasticity in compression is about 2% greater than in tension. Source data are in English units; metric values are converted and rounded.

356.0-16 (continued)

Temperature		Time at temperature, h	Tensile strength		Yield strength		Elongation in 4D, %	Modulus of elasticity(a)	
°F	°C		ksi	MPa	ksi	MPa		10^6 psi	GPa
		10,000	4.0	28	3.0	21	70	6.8	47
700	370	0.5	5.0	34	4.0	28	60	4.2	29
		10	4.0	28	3.0	21	64	4.2	29
		100	3.0	21	2.5	17	68	4.2	29
		1,000	2.5	17	2.0	14	72	4.2	29
		10,000	2.5	17	2.0	14	80	4.2	29

(a) The modulus of elasticity in compression is about 2% greater than in tension. Source data are in English units; metric values are converted and rounded.

360.0-F Die casting: typical tensile properties

Temperature		Time at temperature, h	Tensile strength		Yield strength		Elongation in 2 in. (50 mm), %	Modulus of elasticity(a)	
°F	°C		ksi	MPa	ksi	MPa		10^6 psi	GPa
−112	−80	...	50	345	24	165	2	...	...
−18	−28	...	48	330	25	170	2	...	...
75	25	...	47	325	25	170	3	10.3	71
212	100	0.5	44	305	25	170	4	...	...
		10	44	305	25	170	2	...	...
		100	44	305	26	180	2	...	...
		1,000	48	330	32	220	2	...	...
		10,000	49	340	33	230	2	...	...
300	150	0.5	41	285	25	170	4	...	...
		10	43	295	29	200	3	...	...
		100	44	305	29	200	2	...	...
		1,000	36	250	28	195	4	...	...
		10,000	35	240	27	185	4	...	...
350	177	0.5	38	260	24	165	4	...	...
		10	39	270	28	195	4	...	...
		100	34	235	24	165	4	...	...
		1,000	29	200	21	145	6	...	...
		10,000	27	185	19	130	6	...	...
400	205	0.5	34	235	24	165	4	...	...
		10	31	215	24	165	5	...	...
		100	28	195	21	145	6	...	...
		1,000	24	165	16	110	9	...	...
		10,000	22	150	14	95	10	...	...
500	260	0.5	23	160	18	125	9	...	...
		10	21	145	16	110	10	...	...
		100	18	125	13	90	13	...	...

(continued)

(a) Average of tensile and compressive moduli. Source data are in English units; metric values are converted and rounded.

360.0-F Die casting (continued)

Temperature		Time at temperature, h	Tensile strength		Yield strength		Elongation in 2 in. (50 mm), %	Modulus of elasticity(a)	
°F	°C		ksi	MPa	ksi	MPa		10^6 psi	GPa
		1,000	14	95	9.0	62	17	...	...
		10,000	11	75	7.0	48	22	...	...
600	315	0.5	13	90	9.0	62	14	...	...
		10	12	85	8.0	55	17	...	...
		100	9.0	62	6.0	41	30	...	...
		1,000	7.0	48	4.4	30	45	...	...
		10,000	7.0	48	4.4	30	45	...	...
700	370	0.5	6.0	41	3.8	26	32	...	...
		10	4.5	31	2.9	20	34	...	...
		100	4.4	30	2.8	19	40	...	...
		1,000	4.4	30	2.8	19	40	...	...
		10,000	4.4	30	2.8	19	40	...	...

(a) Average of tensile and compressive moduli. Source data are in English units; metric values are converted and rounded.

380.0-F Die casting: typical tensile properties

Temperature		Time at temperature, h	Tensile strength		Yield strength		Elongation in 2 in. (50 mm), %
°F	°C		ksi	MPa	ksi	MPa	
−320	−196	...	59	405	30	205	2.5
−112	−80	...	49	340	22	150	2.5
−18	−28	...	49	340	23	160	3
75	25	...	48	330	24	165	3
212	100	0.5	46	315	24	165	3
		10	46	315	24	165	3
		100	46	315	24	165	3
		1,000	48	330	27	185	3
		10,000	47	325	29	200	3
300	150	0.5	40	275	23	160	4
		10	42	290	24	165	4
		100	38	260	25	170	4
		1,000	36	250	24	165	4
		10,000	34	235	23	160	4
350	177	0.5	36	250	22	150	4
		10	37	255	23	160	4
		100	33	230	22	150	4
		1,000	32	220	21	145	4
		10,000	31	215	19	130	5
400	205	0.5	32	220	21	145	5
		10	29	200	20	140	5
		100	28	195	19	130	6
		1,000	27	185	17	115	6
		10,000	26	180	16	110	8
500	260	0.5	22	150	15	105	11
		10	21	145	14	95	11
		100	21	145	13	90	12
		1,000	17	115	10	70	18
		10,000	12	83	8.5	59	20
600	315	0.5	13	90	9.0	62	20
		10	12	85	8.0	55	24
		100	9.5	66	6.5	45	27
		1,000	7.0	48	4.6	32	27
		10,000	7.0	48	3.8	26	28
700	370	0.5	6.5	45	4.0	28	28
		10	5.5	38	3.2	22	29
		100	4.5	31	2.5	17	30
		1,000	4.5	31	2.5	17	30
		10,000	4.5	31	2.5	17	30

Source data are in English units; metric values are converted and rounded.

8443.0-F Sand casting: typical tensile properties

Temperature		Time at temperature, h	Tensile strength		Yield strength		Elongation in 4D, %	Modulus of elasticity(a)	
°F	°C		ksi	MPa	ksi	MPa		10^6 psi	GPa
−320	−196	...	25	170	10	70	...	...	...
−112	−80	...	19	130	8.0	55	...	...	...
−18	−28	...	19	130	8.0	55	...	...	...
75	25	...	19	130	8.0	55	8	10.3	71
212	100	0.5	16	110	8.0	55	12	...	...
		10	16	110	8.0	55	12	...	...
		100	16	110	8.0	55	10	...	...
		1,000	16	110	8.5	59	8	...	...
		10,000	16	110	11	75	7	...	...
300	150	0.5	14	95	8.0	55	22	...	...
		10	14	95	9.0	62	21	...	...
		100	14	95	10	70	20	...	...
		1,000	14	95	10	70	19	...	...
		10,000	14	95	9.0	62	18	...	...
400	205	0.5	11	75	7.5	52	25	...	...
		10	11	75	8.5	59	24	...	...
		100	11	75	8.0	55	25	...	...
		1,000	11	75	7.5	52	25	...	...
		10,000	11	75	7.5	52	28	...	...
500	260	0.5	9.0	62	6.0	41	30	...	...
		10	9.0	62	6.0	41	30	...	...
		100	9.0	62	6.0	41	30	...	...
		1,000	9.0	62	6.0	41	30	...	...
		10,000	9.0	62	6.0	41	30	...	...
600	315	0.5	6.5	45	4.8	33	30	...	...
		10	6.5	45	4.8	33	32	...	...
		100	6.5	45	4.8	33	35	...	...
		1,000	5.5	38	4.2	29	40	...	...
		10,000	4.0	28	2.8	19	50	...	...
700	370	0.5	4.7	32	3.4	23	35	...	...
		10	3.8	26	2.6	18	40	...	...
		100	3.4	23	2.3	16	55	...	...
		1,000	3.2	22	2.3	16	55	...	...
		10,000	3.2	22	2.3	16	55	...	...

(a) Average of tensile and compressive moduli. Source data are in English units; metric values are converted and rounded.

520.0-T4 Sand casting: typical tensile properties

Temperature		Time at temperature, h	Tensile strength		Yield strength		Elongation in 4D, %	Modulus of elasticity(a)	
°F	°C		ksi	MPa	ksi	MPa		10^6 psi	GPa
75	25	...	48	330	26	180	16	9.5	66
212	100	0.5	42	290	24	165	...	...	...
		10	43	295	24	165	...	...	...
		100	44	305	25	170	...	...	...
		1,000	44	305	25	170	...	...	...
		10,000	44	305	25	170	...	...	...
300	150	0.5	36	250	22	150	13	...	...
		10	38	260	23	160	14	...	...
		100	39	270	23	160	10	...	...
		1,000	37	255	20	140	13	...	...
		6,500	...	...	...	...	...	...	...
		10,000	37	255	20	140	13	...	...
400	205	0.5	29	200	20	140	11	...	...
		10	27	185	15	105	25	...	...
		100	24	165	12	85	30	...	...
		1,000	22	150	12	85	34	...	...
		2,000	...	...	...	...	...	...	...
		10,000	21	145	12	85	35	...	...
500	260	0.5	21	145	7.0	48	20	...	...
		10	16	110	7.0	48	40	...	...
		100	15	105	7.0	48	42	...	...
		1,000	15	105	7.0	48	48	...	...
		3,000	...	...	...	...	...	...	...
		10,000	15	105	7.0	48	50	...	...
600	315	0.5	11	75	4.0	28	28	...	...
		10	11	75	4.0	28	48	...	...
		100	11	75	4.0	28	52	...	...
		1,000	11	75	4.0	28	60	...	...
		10,000	11	75	4.0	28	60	...	...
700	370	0.5	6.5	45	2.0	14	50	...	...
		10	6.5	45	2.0	14	55	...	...
		100	6.5	45	2.0	14	60	...	...
		1,000	6.5	45	2.0	14	70	...	...
		10,000	6.5	45	2.0	14	70	...	...

(a) Average of tensile and compressive moduli. Source data are in English units; metric values are converted and rounded.

710.0-F Sand casting: typical tensile properties

Temperature		Time at temperature, h	Tensile strength		Yield strength		Elongation in 4D, %
°F	°C		ksi	MPa	ksi	MPa	
−320	−196	...	42(a)	290(a)	35(a)	240(a)	4(a)
−112	−80	...	37(a)	255(a)	28(a)	195(a)	5(a)
−18	−28	...	37(a)	255(a)	26(a)	180(a)	6(a)
75	25	...	35(a)	240(a)	25(a)	170(a)	5(a)
212	100	0.5	34	235	25	170	5
		10	35	240	30	205	5
		100	43	295	38	260	4
		1,000	46	315	42	290	3
		10,000	43	295	40	275	3
300	150	0.5	33	230	25	170	5
		10	37	255	33	230	3
		100	34	235	32	220	3
		1,000	28	195	26	180	6
		10,000	20	140	17	115	12
400	205	0.5	24	165	20	140	10
		10	19	130	17	115	10
		100	15	105	14	95	14
		1,000	12	85	10	70	22
		10,000	10	70	8.0	55	30
500	260	0.5	13	90	11	75	12
		10	10	70	8.5	59	23
		100	8.0	55	7.0	48	33
		1,000	7.0	48	6.0	41	45
		10,000	6.5	45	5.0	34	60

(a) 30 days after casting. Source data are in English units; metric values are converted and rounded.

APPENDIX 2

Physical Properties of Aluminum and Aluminum Alloys

Table A2.1 Physical properties of wrought aluminum alloys—engineering units

Alloy	Average(a) coefficient of thermal expansion, 68 °F to 212 °F per °F	Melting range(b)(c) approx., °F	Temper	Thermal conductivity at 77 °F, Btu · in./h · ft² · °F	Electrical conductivity at 68 °F percent of international annealed copper standard: Equal volume	Electrical conductivity at 68 °F percent of international annealed copper standard: Equal weight	Electrical resistivity at 68 °F, Ω circular-mil/ft
1060	13.1	1195–1215	O	1625	62	204	17
			H18	1600	61	201	17
1100	13.1	1190–1215	O	1540	59	194	18
			H18	1510	57	187	18
1350	13.2	1195–1215	All	1625	62	204	17
2011	12.7	1005–1190(d)	T3	1050	39	123	27
			T8	1190	45	142	23
2014	12.8	945–1180(d)	O	1340	50	159	21
			T4	930	34	108	31
			T6	1070	40	127	26
2017	13.1	955–1185(d)	O	1340	50	159	21
			T4	930	34	108	31
2018	12.4	945–1180(e)	T61	1070	40	127	26
2024	12.9	935–1180(d)	O	1340	50	160	21

(continued)

(a) Coefficient to be multiplied by 10–6. Example 12.2 × 10–6 = 0.0000122. (b) Melting ranges shown apply to wrought products of ¼ inch thickness or greater. (c) Based on typical composition of the indicated alloys. (d) Eutectic melting is not eliminated by homogenization. (e) Eutectic melting can be completely eliminated by homogenization. (f) Homogenization may raise eutectic melting temperature 20–40 °F but usually does not eliminate eutectic melting. (g) Although not formerly registered, the literature and some specifications have used T736 as the designation for this temper. Source: Ref A2.1. Adapted with permission from Aluminum Association

Table A2.1 (continued)

Alloy	Average(a) coefficient of thermal expansion 68 °F to 212 °F per °F	Melting range(b)(c) approx. °F	Temper	Thermal conductivity at 77 °F Btu · in./h · ft² · °F	Electrical conductivity at 68 °F percent of international annealed copper standard: Equal volume	Equal weight	Electrical resistivity at 68 °F Ω circular-mil/ft
2024	12.9	935–1180(d)	T3, T4, T361	840	30	96	35
			T6, T81, T861	1050	38	122	27
2025	12.6	970–1185(d)	T6	1070	40	128	26
2036	13.0	1030–1200(e)	T4	1100	41	135	25
2117	13.2	1030–1200(e)	T4	1070	40	130	26
2124	12.7	935–1180(d)	T851	1055	38	122	27
2218	12.4	940–1175(d)	T72	1070	40	126	26
2219	12.4	1010–1190(e)	O	1190	44	138	24
			T31, T37	780	28	88	37
			T6, T81, T87	840	30	94	35
2618	12.4	1020–1180	T6	1020	37	120	28
3003	12.9	1190–1210	O	1340	50	163	21
			H12	1130	42	137	25
			H14	1100	41	134	25
			H18	1070	40	130	26
3004	13.3	1165–1210	All	1130	42	137	25
3105	13.1	1175–1210	All	1190	45	148	23
4032	10.8	990–1060(d)	O	1070	40	132	26
			T6	960	35	116	30
4043	12.3	1065–1170	O	1130	42	140	25
4045	11.7	1065–1110	All	1190	45	151	23
4343	12.0	1070–1135	All	1250	47	158	25
5005	13.2	1170–1210	All	1390	52	172	20
5050	13.2	1155–1205	All	1340	50	165	21
5052	13.2	1125–1200	All	960	35	116	30
5056	13.4	1055–1180	O	810	29	98	36
			H38	750	27	91	38
5083	13.2	1095–1180	O	810	29	98	36
5086	13.2	1085–1185	All	870	31	104	33
5154	13.3	1100–1190	All	870	32	107	32
5252	13.2	1125–1200	All	960	35	116	30
5254	13.3	1100–1190	All	870	32	107	32
5356	13.4	1060–1175	O	810	29	98	36
5454	13.1	1115–1195	O	930	34	113	31
			H38	930	34	113	31

(continued)

(a) Coefficient to be multiplied by 10–6. Example 12.2 × 10–6 = 0.0000122. (b) Melting ranges shown apply to wrought products of ¼ inch thickness or greater. (c) Based on typical composition of the indicated alloys. (d) Eutectic melting is not eliminated by homogenization. (e) Eutectic melting can be completely eliminated by homogenization. (f) Homogenization may raise eutectic melting temperature 20–40 °F but usually does not eliminate eutectic melting. (g) Although not formerly registered, the literature and some specifications have used T736 as the designation for this temper. Source: Ref A2.1. Adapted with permission from Aluminum Association

Table A2.1 (continued)

Alloy	Average(a) coefficient of thermal expansion, 68 °F to 212 °F per °F	Melting range(b)(c) approx., °F	Temper	Thermal conductivity at 77 °F, Btu · in./h · ft² · °F	Electrical conductivity at 68 °F percent of international annealed copper standard, Equal volume	Equal weight	Electrical resistivity at 68 °F, Ω circular-mil/ft
5456	13.3	1055–1180	O	810	29	98	36
5457	13.2	1165–1210	All	1220	46	153	23
5657	13.2	1180–1215	All	1420	54	180	19
6005	13.0	1125–1210(e)	T1	1250	47	155	22
			T5	1310	49	161	21
6005A	13.0	1110–1200	T1	1220	47	155	22
6005A	13.0	1110–1200	T5	1340	50	165	21
6005A	13.0	1110–1200	T61	1310	49	161	21
6053	12.8	1070–1205(e)	O	1190	45	148	23
			T4	1070	40	132	26
			T6	1130	42	139	25
6061	13.1	1080–1205(e)	O	1250	47	155	22
			T4	1070	40	132	26
			T6	1160	43	142	24
6063	13.0	1140–1210	O	1510	58	191	18
			T1	1340	50	165	21
			T5	1450	55	181	19
			T6, T83	1390	53	175	20
6066	12.9	1045–1195(e)	O	1070	40	132	26
			T6	1020	37	122	28
6070	...	1050–1200(e)	T6	1190	44	145	24
6082	12.8	1070–1200	T6, T6511	1190	44	145	24
6101	13.0	1150–1210	T6	1510	57	188	18
			T61	1540	59	194	18
			T63	1510	58	191	18
			T64	1570	60	198	17
			T65	1510	58	191	18
6105	13.0	1110–1200(e)	T1	1220	46	151	23
			T5	1340	50	165	21
6151	12.9	1090–1200(e)	O	1420	54	178	19
			T4	1130	42	138	25
			T6	1190	45	148	23
6201	13.0	1125–1210(e)	T81	1420	54	180	19
6262	13.0	1080–1205(e)	T9	1190	44	145	24
6351	13.0	1030–1200	T6	1220	46	151	23

(continued)

(a) Coefficient to be multiplied by 10–6. Example 12.2 × 10–6 = 0.0000122. (b) Melting ranges shown apply to wrought products of ¼ inch thickness or greater. (c) Based on typical composition of the indicated alloys. (d) Eutectic melting is not eliminated by homogenization. (e) Eutectic melting can be completely eliminated by homogenization. (f) Homogenization may raise eutectic melting temperature 20–40 °F but usually does not eliminate eutectic melting. (g) Although not formerly registered, the literature and some specifications have used T736 as the designation for this temper. Source: Ref A2.1. Adapted with permission from Aluminum Association

Table A2.1 (continued)

Alloy	Average(a) coefficient of thermal expansion 68 °F to 212 °F per °F	Melting range(b)(c) approx. °F	Temper	Thermal conductivity at 77 °F Btu · in./h · ft² · °F	Electrical conductivity at 68 °F percent of international annealed copper standard: Equal volume	Equal weight	Electrical resistivity at 68 °F Ω circular-mil/ft
6360	13.0	1140–1210	T5	1300	52	172	20
	13.0		T6	1320	53	175	20
6463	13.0	1140–1210(e)	T1	1340	50	165	21
			T5	1450	55	181	19
			T6	1390	53	175	20
6951	13.0	1140–1210	O	1480	56	186	19
			T6	1370	52	172	20
7049	13.0	890–1175	T73	1070	40	132	26
7050	12.8	910–1165	T74(g)	1090	41	135	25
7072	13.1	1185–1215	O	1540	59	193	18
7075	13.1	890–1175(f)	T6	900	33	105	31
7175	13.0	890–1175(f)	T74(g)	1080	39	124	26
7475	12.9	890–1175	T61, T651	960	35	116	30
			T76, T761	1020	40	132	26
			T7351	1130	42	139	25
8017	13.1	1190–1215	H12, H22	...	59	193	18
			H212	...	61	200	17
8030	13.1	1190–1215	H221	1600	61	201	17
8176	13.1	1190–1215	H24	...	61	201	17

(a) Coefficient to be multiplied by 10–6. Example 12.2 × 10–6 = 0.0000122. (b) Melting ranges shown apply to wrought products of ¼ inch thickness or greater. (c) Based on typical composition of the indicated alloys. (d) Eutectic melting is not eliminated by homogenization. (e) Eutectic melting can be completely eliminated by homogenization. (f) Homogenization may raise eutectic melting temperature 20–40 °F but usually does not eliminate eutectic melting. (g) Although not formerly registered, the literature and some specifications have used T736 as the designation for this temper. Source: Ref A2.1. Adapted with permission from Aluminum Association

Table A2.2 Physical properties of wrought aluminum alloys—metric SI units

Alloy	Average(a) coefficient of thermal expansion	Melting range(b)(c) approx.	Temper	Thermal conductivity at 25 °C	Electrical conductivity at 20 °C Ms/m(h)		Electrical resistivity at 20 °C
	20 to 100 °C per °C	°C		W/mK	Equal volume	Equal mass	Ω mm²/m
1060	23.6	645–655	O	234	36	118	0.028
			H18	230	35	117	0.029
1100	23.6	640–655	O	222	34	113	0.029
			H18	218	33	108	0.030
1350	23.6	645–655	All	234	36	118	0.028
2011	22.9	540–645(e)	T3	151	23	71	0.043
			T8	172	26	82	0.038
2014	23.0	505–635(d)	O	193	29	92	0.034
			T4	134	20	63	0.050
			T6	155	23	74	0.043
2017	23.6	510–640(d)	O	193	29	92	0.034
			T4	134	20	63	0.050
2018	22.3	505–640(e)	T61	155	23	74	0.043
2024	23.2	500–635(d)	O	193	29	93	0.034
			T3, T4, T361	121	17	56	0.059
			T6, T81, T861	151	22	71	0.045
2025	22.7	520–640(d)	T6	155	23	74	0.043
2036	23.4	555–650(e)	T4	159	24	78	0.042
2117	23.8	555–650(e)	T4	155	23	75	0.043
2124	22.9	500–635(d)	T851	152	22	71	0.045
2218	22.3	505–635(d)	T72	155	23	73	0.043
2219	22.3	545–645(d)	O	172	26	80	0.038
			T31, T37	113	16	57	0.062
			T6, T81, T87	121	177	58	0.059
2618	22.3	550–640	T6	146	21	70	0.048
3003	23.2	640–655	O	193	29	92	0.064
			H12	163	24	78	0.042
			H14	159	24	78	0.042
			H18	155	23	74	0.043
3004	23.9	630–655	All	163	24	79	0.042
3105	23.6	635–655	All	172	26	86	0.038
4032	19.4	530–570(d)	O	155	23	77	0.043
			T6	138	20	67	0.050
4043	22.0	575–630	O	163	24	81	0.041
4045	21.1	575–600	All	171	26	88	0.038

(continued)

(a) Coefficient to be multiplied by 10–6. Example 23.6 × 10–6 = 0.0000236. (b) Melting ranges shown apply to wrought products of 6 mm thickness or greater (c) Based on typical composition of the indicated alloys (d) Eutectic melting is not eliminated by homogenization (e) Eutectic melting can be completely eliminated by homogenization (f) Homogenization may raise eutectic melting temperature 10–20 °C but usually does not eliminate eutectic melting (g) Although not formerly registered, the literature and some specifications have used T736 as the designation for this temper (h) MS/m = 0.58 % IACS. Source: Ref A2.2. Adapted with permission from Aluminum Association

Table A2.2 (continued)

Alloy	Average(a) coefficient of thermal expansion 20 to 100 °C per °C	Melting range(b)(c) approx. °C	Temper	Thermal conductivity at 25 °C W/mK	Electrical conductivity at 20 °C Ms/m(h) Equal volume	Electrical conductivity at 20 °C Ms/m(h) Equal mass	Electrical resistivity at 20 °C Ω mm²/m
4343	21.6	575–615	All	180	27	92	0.037
5005	23.8	630–655	All	201	30	100	0.033
5050	23.8	625–650	All	193	29	96	0.034
5052	23.8	605–650	All	138	20	67	0.050
5056	24.1	565–640	O	117	17	57	0.059
			H38	109	16	53	0.062
5083	23.8	580–640	O	117	17	57	0.059
5086	23.8	585–640	All	126	18	60	0.056
5154	23.9	590–645	All	126	19	62	0.053
5252	23.8	605–650	All	138	20	67	0.050
5254	23.9	590–645	All	126	19	62	0.053
5356	24.1	575–635	O	117	17	57	0.059
5454	23.6	600–645	O	134	20	66	0.050
			H38	137	20	66	0.050
5456	23.9	570–640	O	117	17	57	0.059
5457	23.8	630–655	All	176	27	89	0.037
5652	23.8	605–650	All	138	20	69	0.050
5657	23.8	635–655	All	205	31	104	0.032
6005	23.6	605–655(e)	T1	180	27	90	0.037
			T5	188	28	93	0.036
6005A	23.6	600–650	T1	176	27	90	0.037
6005A	23.6	600–650	T5	193	29	96	0.034
6005A	23.6	600–650	T61	188	28	93	0.036
6053	23.0	575–650(e)	O	172	26	86	0.038
			T4	155	23	77	0.042
			T6	167	24	81	0.041
6061	23.6	580–650(e)	O	180	27	90	0.037
			T4	155	23	77	0.043
			T6	167	25	82	0.040
6063	23.4	615–655	O	218	34	111	0.029
			T1	193	29	96	0.034
			T5	209	32	105	0.031
			T6, T83	201	31	102	0.032
6066	23.2	560–645(d)	O	155	23	77	0.043
			T6	146	21	71	0.048

(continued)

(a) Coefficient to be multiplied by 10–6. Example 23.6 × 10–6 = 0.0000236. (b) Melting ranges shown apply to wrought products of 6 mm thickness or greater (c) Based on typical composition of the indicated alloys (d) Eutectic melting is not eliminated by homogenization (e) Eutectic melting can be completely eliminated by homogenization (f) Homogenization may raise eutectic melting temperature 10–20 °C but usually does not eliminate eutectic melting (g) Although not formerly registered, the literature and some specifications have used T736 as the designation for this temper (h) MS/m = 0.58 % IACS. Source: Ref A2.2. Adapted with permission from Aluminum Association

Table A2.2 (continued)

Alloy	Average(a) coefficient of thermal expansion 20 to 100 °C per °C	Melting range(b)(c) approx. °C	Temper	Thermal conductivity at 25 °C W/mK	Electrical conductivity at 20 °C Ms/m(h) Equal volume	Equal mass	Electrical resistivity at 20 °C Ω mm²/m
6070	...	565–650(d)	T6	172	26	84	0.038
6082	23.0	575–650	T6, T6511	172	26	84	0.040
6101	23.4	620–655	T6	218	33	109	0.030
			T61	222	34	113	0.029
			T63	218	34	111	0.029
			T64	226	35	115	0.029
			T65	218	34	111	0.029
6105	23.4	600–650(e)	T1	176	27	88	0.037
			T5	183	29	96	0.034
6151	23.2	590–650(e)	O	205	31	103	0.032
			T4	163	24	80	0.042
			T6	172	26	86	0.038
6201	23.4	610–655(e)	T81	205	31	104	0.032
6262	23.4	580–650(e)	T9	172	26	84	0.038
6351	23.4	555–650	T6	176	27	88	0.038
6463	23.4	615–655(e)	T1	193	29	96	0.034
			T5	209	32	105	0.031
			T6	201	31	102	0.032
6951	23.4	615–655	O	213	32	108	0.031
		475–635	T6	197	30	100	0.033
7049	23.4	490–630	T73	155	23	77	0.043
7050	23.0	640–655	T74(h)	1577	24	78	0.042
7072	23.6	475–635(f)	O	222	34	112	0.029
7075	23.6	475–635(f)	T6	130	19	61	0.053
7175	23.4	475–630(f)	T74	157	23	72	0.043
7178	23.4	475–635	T6	126	18	57	0.056
7475	23.2		T61, T651	138	20	69	0.050
			T76, T761	146	23	77	0.043
			T7351	163	24	81	0.041
8017	23.6	645–655	H12, H22	...	34	113	0.029
	23.6		H212	...	35	117	0.029
8030	23.6	645–655	H221	230	35	117	0.029
8176		645–655	H24	230	35	117	0.029

(a) Coefficient to be multiplied by 10–6. Example 23.6 × 10–6 = 0.0000236. (b) Melting ranges shown apply to wrought products of 6 mm thickness or greater (c) Based on typical composition of the indicated alloys (d) Eutectic melting is not eliminated by homogenization (e) Eutectic melting can be completely eliminated by homogenization (f) Homogenization may raise eutectic melting temperature 10–20 °C but usually does not eliminate eutectic melting (g) Although not formerly registered, the literature and some specifications have used T736 as the designation for this temper (h) MS/m = 0.58 % IACS. Source: Ref A2.2. Adapted with permission from Aluminum Association

Table A2.3 Typical physical properties of cast aluminum alloys(a)(b)

			Density		Approximate melting range(c)		Thermal conductivity		Electrical conductivity				Electrical resistivity		Specific heat	
							At 77 °F	At 25 °C	At 68 °F(d) Volume	At 68 °F(d) Weight	At 25 °C Volume	At 25 °C Weight	At 68 °F	At 20 °C	At 68 °F	At 20 °C
Type of casting	Alloy	Temper	lb/in³	g/cm³	°F	°C	Btu · in./h · ft² · °F	W/m · K	%IACS	%IACS	MS/m	MS/m	Ω circular–mil/ft	Ω mm²/m	Btu/lb · °F	J/kg · °C
Sand or permanent mold	100.0	F	0.096	2.70	970–1160	520–630	1165	168	54	177	31	103	19	0.032	0.210	879
	201.0	T6	0.101	2.80	1060–1200	570–650	840	121	30	99	17	57	35	0.057	0.220	922
	201.0	T7	0.101	2.80	1060–1200	570–650	840	121	33	108	19	63	32	0.052	0.220	922
	201.0	T43	0.101	2.80	1060–1200	570–650	...	...	...	...	...	...	...	...	0.220	922
	204.0	T4	0.101	2.80	1060–1200	570–650	840	121	29	95	17	55	36	0.059	0.230	963
	208.0	F	0.101	2.80	970–1160	520–630	840	121	31	102	18	59	34	0.056	0.230	963
	222.0	F	0.107	2.95	970–1160	520–630	925	133	34	112	20	65	31	0.051	0.230	963
	222.0	O	0.107	2.95	970–1160	520–630	1095	158	41	135	24	78	25	0.042	0.230	963
	222.0	T61	0.101	2.95	970–1160	520–630	895	129	33	108	19	63	32	0.052	0.230	963
	224.0	T62	0.102	2.81	1020–1190	550–645	810	117	30	99	17	57	35	0.057	0.230	963
	240.0	F	0.100	2.78	950–1110	515–605	665	96	23	76	13	44	45	0.075	0.230	963
	242.0	O	0.102	2.81	990–1175	525–635	925	133	33	108	19	63	32	0.052	0.230	963
	242.0	T571	0.102	2.81	990–1175	525–635	925	133	34	112	20	65	31	0.051	0.230	963
	242.0	T61	0.102	2.81	990–1175	525–635	925	133	33	108	19	63	32	0.052	0.230	963
	242.0	T77	0.102	2.81	990–1175	525–635	1040	150	38	125	22	73	27	0.045	0.230	963
	295.0	T4	0.102	2.81	970–1190	520–645	955	138	35	115	20	67	30	0.049	0.230	963
	295.0	T6	0.102	2.81	970–1190	520–645	955	138	35	115	20	67	30	0.049	0.230	963

(continued)

(a) Taken from the following references: Standards for Aluminum Sand and Permanent Mold Castings, The Aluminum Association, Inc., Dec., 1992; Aluminum Casting Technology, 2nd Edition, The American Foundrymen's Society, Inc., 1993; Product Design for Diecasting, Diecasting Development Council; A. Kearny and E.L. Rooy, Aluminum Foundry Products, ASM Metals Handbook, Vol. 2, 1998, p 123–177. (b) Metric unit values generally derived from engineering/English unit values. (c) Melting ranges based upon nominal composition of each alloy, in thickness of 1/4 in. (6 mm) or more. (d) %IACS = Percent of International Annealed Copper Standard. Source: Ref A2.3

Table A2.3 (continued)

									Electrical conductivity							
					Approximate melting range(c)		Thermal conductivity		At 68 °F(d)		At 25 °C		Electrical resistivity		Specific heat	
			Density				At 77 °F	At 25 °C	Volume	Weight	Volume	Weight	At 68 °F	At 20 °C	At 68 °F	At 20 °C
Type of casting	Alloy	Temper	lb/in^3	g/cm^3	°F	°C	Btu · in./h · ft^2 · °F	W/m · K	%IACS	%IACS	MS/m	MS/m	Ω circular–mil/ft	Ω mm^2/m	Btu/lb · °F	J/kg · °C
	295.0	T62	0.102	2.81	970–1190	520–645	980	141	35	115	20	67	30	0.049	0.230	963
	319.0	F	0.101	2.79	960–1120	520–605	780	112	28	92	16	53	37	0.061	0.230	963
	328.0	F	0.098	2.70	1025–1105	555–595	665	96	30	99	17	57	35	0.057	0.230	963
	333.0	F	0.100	2.77	960–1085	520–585	720	104	26	85	15	50	40	0.066	0.230	963
	333.0	T5	0.100	2.77	960–1085	520–585	810	117	29	95	17	55	36	0.059	0.230	963
	333.0	T6	0.100	2.77	960–1085	520–585	810	117	29	95	17	55	36	0.059	0.230	963
	333.0	T7	0.100	2.77	960–1085	520–585	955	138	35	115	20	67	30	0.049	0.230	963
	355.0	T51	0.098	2.71	1015–1150	550–620	1155	166	43	141	25	82	24	0.040	0.230	963
	355.0	T6	0.098	2.71	1015–1150	550–620	1010	145	38	125	22	73	27	0.045	0.230	963
	355.0	T61	0.098	2.71	1015–1150	550–620	1010	145	37	122	21	71	28	0.047	0.230	963
	355.0	T62	0.098	2.71	1015–1150	550–620	1010	145	38	125	22	73	27	0.045	0.230	963
	355.0	T7	0.098	2.71	1015–1150	550–620	1125	162	42	138	24	80	25	0.041	0.230	963
	355.0	T71	0.098	2.71	1015–1150	550–620	1040	150	39	128	23	74	27	0.044	0.230	963
	C355.0	T6	0.098	2.71	1015–1150	550–620	1040	150	39	128	23	74	27	0.044	0.230	963
	C355.0	T61	0.098	2.71	1015–1150	550–620	1010	145	43	141	25	82	24	0.040	0.230	963
	356.0	T51	0.097	2.68	1035–1135	560–615	1155	166	43	141	25	82	24	0.040	0.230	963
	356.0	T6	0.097	2.68	1035–1135	560–615	1040	150	40	131	23	76	26	0.043	0.230	963
	356.0	T7	0.097	2.68	1035–1135	560–615	1110	160	42	138	24	80	25	0.041	0.230	963

(continued)

(a) Taken from the following references: Standards for Aluminum Sand and Permanent Mold Castings, The Aluminum Association, Inc., Dec., 1992; Aluminum Casting Technology, 2nd Edition, The American Foundrymen's Society, Inc., 1993; Product Design for Diecasting, Diecasting Development Council; A. Kearny and E.L. Rooy, Aluminum Foundry Products, ASM Metals Handbook, Vol. 2, 1998, p 123–177. (b) Metric unit values generally derived from engineering/English unit values. (c) Melting ranges based upon nominal composition of each alloy, in thickness of 1/4 in. (6 mm) or more. (d) %IACS = Percent of International Annealed Copper Standard. Source: Ref A2.3

Table A2.3 (continued)

			Density		Approximate melting range(c)		Thermal conductivity		Electrical conductivity				Electrical resistivity		Specific heat	
									At 68 °F(d)		At 25 °C					
							At 77 °F	At 25 °C	Volume	Weight	Volume	Weight	At 68 °F	At 20 °C	At 68 °F	At 20 °C
Type of casting	Alloy	Temper	lb/in³	g/cm³	°F	°C	Btu · in./h · ft² · °F	W/m · K	%IACS	%IACS	MS/m	MS/m	Ω circular–mil/ft	Ω mm²/m	Btu/lb · °F	J/kg · °C
	A356.0	T6	0.097	2.67	1035–1135	560–610	1040	150	40	131	23	76	26	0.043	0.230	963
	A356.0	T61	0.097	2.67	1035–1135	560–610	1040	150	39	128	23	74	27	0.044	0.230	963
	357.0	F	0.097	2.67	1035–1135	560–615	1040	150	39	128	23	74	27	0.044	0.230	963
	357.0	T6	0.097	2.67	1035–1135	560–615	1040	150	39	128	23	74	27	0.044	0.230	963
	A357.0	T6	0.097	2.67	1035–1135	555–610	1100	158	40	131	23	76	26	0.043	0.230	963
	A357.0	T61	0.097	2.67	1035–1135	555–610	1040	150	39	128	23	74	27	0.044	0.230	963
	359.0	T6	0.097	2.67	1045–1115	565–600	955	138	35	115	20	67	30	0.049	0.230	963
	443.0	F	0.097	2.69	1065–1170	575–630	1010	145	37	122	21	71	28	0.047	0.230	963
	443.0	O	0.097	2.69	1065–1170	575–630	1125	162	42	138	24	80	25	0.041	0.230	963
	B443.0	F	0.097	2.69	1065–1170	575–630	1010	145	37	122	21	71	28	0.047	0.230	963
	A444.0	F	0.097	2.68	1070–1170	575–630	1095	158	41	135	24	78	25	0.042	0.230	963
	511.0	F	0.100	2.66	1090–1185	590–640	980	141	36	118	21	69	29	0.048	0.230	963
	512.0	F	0.096	2.65	1090–1170	590–630	1010	145	38	125	22	73	27	0.045	0.230	963
	514.0	F	0.096	2.65	1110–1185	600–640	954	137	35	115	20	67	30	0.049	0.230	963
	520.0	T4	0.093	2.57	840–1120	450–600	605	87	21	69	12	40	50	0.082	0.230	963
	535.0	F	0.095	2.62	1020–1165	550–630	695	100	23	76	13	44	45	0.075	0.230	963

(continued)

(a) Taken from the following references: Standards for Aluminum Sand and Permanent Mold Castings, The Aluminum Association, Inc., Dec., 1992; Aluminum Casting Technology, 2nd Edition, The American Foundrymen's Society, Inc., 1993; Product Design for Diecasting, Diecasting Development Council; A. Kearny and E.L. Rooy, Aluminum Foundry Products, ASM Metals Handbook, Vol. 2, 1998, p 123–177. (b) Metric unit values generally derived from engineering/English unit values. (c) Melting ranges based upon nominal composition of each alloy, in thickness of 1/4 in. (6 mm) or more. (d) %IACS = Percent of International Annealed Copper Standard. Source: Ref A2.3

Table A2.3 (continued)

			Density		Approximate melting range(c)		Thermal conductivity		Electrical conductivity				Electrical resistivity		Specific heat	
									At 68 °F(d)		At 25 °C					
							At 77 °F	At 25 °C	Volume	Weight	Volume	Weight	At 68 °F	At 20 °C	At 68 °F	At 20 °C
Type of casting	Alloy	Temper	lb/in³	g/cm³	°F	°C	Btu · in./h · ft² · °F	W/m · K	%IACS	%IACS	MS/m	MS/m	Ω circular-mil/ft	Ω mm²/m	Btu/lb · °F	J/kg · °C
	A535.0	F	0.090	2.54	1020–1150	550–620	695	100	23	76	13	44	45	0.075	0.230	963
	705.0	F	0.100	2.76	1105–1180	600–640	720	104	25	79	15	46	43	0.071	0.230	963
	707.0	F	0.100	2.77	1085–1165	585–630	720	104	25	79	15	46	43	0.071	0.230	963
	710.0	F	0.102	2.81	1105–1195	600–650	955	138	35	111	20	64	31	0.051	0.230	963
	712.0	F	0.101	2.81	1135–1200	600–640	1100	158	35	111	20	64	31	0.051	0.230	963
	713.0	F	0.100	2.81	1100–1180	595–630	1070	154	30	95	17	55	36	0.059	0.230	963
	771.0	F	0.102	2.81	1120–1190	550–645	955	138	37	117	21	68	29	0.048	0.230	963
	850.0	T5	0.104	2.88	435–1200	225–650	1290	186	47	154	27	90	22	0.037	0.230	963
	851.0	T5	0.103	2.83	440–1165	230–630	1155	166	43	141	25	82	24	0.040	0.230	963
	852.0	T5	0.104	2.88	400–1175	210–635	1215	175	45	148	26	86	23	0.038	0.230	963
Die	360.0	F	0.095	2.63	1035–1105	557–596	785	113	30	99	17	57	35	0.057	0.230	963
	A360.0	F	0.095	2.63	1035–1105	557–596	785	113	29	95	17	55	36	0.059	0.230	963
	380.0	F	0.099	2.74	1000–1100	540–595	667	96	27	89	16	52	39	0.064	0.230	963
	A380.0	F	0.098	2.71	1000–1100	540–595	667	96	23	76	13	44	45	0.075	0.230	963

(continued)

(a) Taken from the following references: Standards for Aluminum Sand and Permanent Mold Castings, The Aluminum Association, Inc., Dec., 1992; Aluminum Casting Technology, 2nd Edition, The American Foundrymen's Society, Inc., 1993; Product Design for Diecasting, Diecasting Development Council; A. Kearny and E.L. Rooy, Aluminum Foundry Products, ASM Metals Handbook, Vol. 2, 1998, p 123–177. (b) Metric unit values generally derived from engineering/English unit values. (c) Melting ranges based upon nominal composition of each alloy, in thickness of 1/4 in. (6 mm) or more. (d) %IACS = Percent of International Annealed Copper Standard. Source: Ref A2.3

Table A2.3 (continued)

			Density		Approximate melting range(c)		Thermal conductivity		Electrical conductivity				Electrical resistivity		Specific heat	
									At 68 °F(d)		At 25 °C					
							At 77 °F	At 25 °C	Volume	Weight	Volume	Weight	At 68 °F	At 20 °C	At 68 °F	At 20 °C
Type of casting	Alloy	Temper	lb/in^3	g/cm^3	°F	°C	Btu · in./h · ft^2 · °F	W/m · K	%IACS	%IACS	MS/m	MS/m	Ω circular–mil/ft	Ω mm^2/m	Btu/lb · °F	J/kg · °C
	383.0	F	0.099	2.74	960–1080	516–582	667	96	23	76	13	44	45	0.075	0.230	963
	384.0	F	0.102	2.82	960–1080	516–582	667	96	22	72	13	42	47	0.078	...	...
	B390.0	F	0.098	2.73	950–1200	510–650	930	134	27	89	16	52	39	0.064	...	...
	413.0	F	0.096	2.66	1065–1080	574–582	840	121	31	102	18	59	34	0.056	0.230	963
	A413.0	F	0.096	2.66	1065–1080	574–582	840	121	31	102	18	59	34	0.056	0.230	963
	C443.0	F	0.097	2.69	1065–1170	574–632	985	142	37	122	21	71	28	0.047	0.230	963
	518.0	F	0.093	2.57	995–1150	535–621	667	96	24	79	14	46	43	0.072	...	...

(a) Taken from the following references: Standards for Aluminum Sand and Permanent Mold Castings, The Aluminum Association, Inc., Dec., 1992; Aluminum Casting Technology, 2nd Edition, The American Foundrymen's Society, Inc., 1993; Product Design for Diecasting, Diecasting Development Council; A. Kearny and E.L. Rooy, Aluminum Foundry Products, ASM Metals Handbook, Vol. 2, 1998, p 123-177. (b) Metric unit values generally derived from engineering/English unit values. (c) Melting ranges based upon nominal composition of each alloy, in thickness of 1/4 in. (6 mm) or more. (d) %IACS = Percent of International Annealed Copper Standard. Source: Ref A2.3

REFERENCES

A2.1 *Aluminum Standards and Data 2013,* The Aluminum Association, Arlington, VA, 2013

A2.2 *Aluminum Standards and Data 2013 Metric SI,* The Aluminum Association, Arlington, VA, 2013

A2.3 J.G. Kaufman and E.L. Rooy, *Aluminum Alloys Castings: Properties, Processes, and Applications,* ASM International, 2004

APPENDIX 3

Representative Fire Test Reports for Aluminum Alloys

Reproduced with permission from the files of the Aluminum Association, Inc.

Representative fire test reports (Fig. A3.1–A3.8) for aluminum alloys include the following:

- United States Testing Company, Inc. Incombustibility Tests, Report 10107 dated July 19, 1967
- Signet Testing Laboratories, Inc. Noncombustibility Tests, Report dated September 30, 1968
- Signet Testing Laboratories, Inc. Noncombustibility Tests, Report dated May 17, 1972
- Warrington Fire Research Centre Report 46083-CWM/KC dated August 17, 1989
- Warrington Fire Research Centre Report 48322-MT/KC dated February 1, 1990
- Certificate of Inspection & Tests carried out by the Marine Surveyors of the London Department of Transport dated August 31, 1983
- Summary of Fire Insurers' Research & Testing Organization (FIRTO) Report dated October 10, 1983
- Lloyd's Register of Shipping Certificate No. ICD/F83/697

11-25-00 01:06P TEMCOR 310 549 6807 P.04

ESTABLISHED 1880

United States Testing Company, Inc.
HOBOKEN, N. J. 07030
201-792-2400

REPORT

RECEIVED
JUL 24 1967
W. W. FRITSKY

NUMBER
10107
(Refer to this number)

July 19, 1967

Client: The Aluminum Association
420 Lexington Avenue
New York, New York 10017

Subject: Two (2) samples of aluminum alloy, sampled and identified by Client for specification tests. Letter dated 7-11-67.

Sample Identification:*

1. 6061-T6 Aluminum Alloy (1" cubes)
2. 6063-T5 Aluminum Alloy (1" cubes)

Test Performed:

The submitted samples were tested for conformance to the requirements of "Incombustible Material" as defined in Section 410 Volume I of the Uniform Building Code 1961 Edition, published by the International Conference of Building Officials.

Test Procedure:

Testing for conformance to the above stated requirement was performed in accordance with the procedures specified in the Uniform Building Code Standard No. 4-1-61 Volume III-1961 Edition.

*Certification of material as supplied by Client is appended to this report.

Test Results:

	Exposure Temperature, °F. 5 minute duration		
6061-T6	Start	Finish	Observations
Specimen			
1	1205	1200	Conforms
2	1205	1205	Conforms
3	1205	1205	Conforms
6063-T5			
Specimen			
1	1205	1210	Conforms
2	1205	1210	Conforms
3	1210	1210	Conforms

Page 1 of 2

Supervised by John Carroll
Metallurgist

United States Testing Company, Inc.
F. Rezzuoli

Our letters and reports are for the exclusive use of the client to whom they are addressed, and their communication to any others, or the use of the name of United States Testing Company, Inc., must receive our prior written approval. Our letters and reports apply only to the sample tested and are not necessarily indicative of the qualities of apparently identical or similar products. Samples not destroyed in testing are retained a maximum of thirty days. The reports and letters and the name of the United States Testing Company, Inc., or its seals or insignia, are not to be used under any circumstances [illegible]

Fig. A3.1 United States Testing Company, Inc. Incombustibility Tests, Report 10107 dated July 19, 1967, illustrating that aluminum alloys 6061 and 6063 conformed to the requirements of incombustible material as defined by the *Uniform Building Code* published by the International Conference of Building Officials

Jul 25 00 01:06P TEMCOR 310 549 0807 P.03

LABORATORY CERTIFICATE

SIGNET TESTING LABORATORIES, INC.

TESTING AND INSPECTION OF CONSTRUCTIONAL AND INDUSTRIAL MATERIALS

ENGINEERS
CHEMISTS
METALLURGISTS

STL

REPLY TO
1428 WEST WINTON AVENUE
HAYWARD, CALIFORNIA 94545
782-7919 AREA CODE 415

September 30, 1968

P.O. # 90-90404

LAB NO.: 4342
SAMPLES: Two (2) Samples of Aluminum Alloy
MARK: 3004 Alloy and 8112 Alloy
DATE TESTED: August 23, 1968 and September 17, 1968
REPORT TO: Kaiser Aluminum and Chemical Corporation
300 Lakeside Drive
Oakland, California 94604
Attention: Mr. R.A. Edgren

REPORT

TEST

The submitted samples of aluminum alloys were tested for noncombustibility property as prescribed by ASTM Designation: E136-65, "Standard Method of Test for Determining Noncombustibility of Elementary Materials".

TEST SPECIMENS AND PROCEDURES

The prepared test specimens were 1½" wide by 1½" thick by 2" long. The test was performed in accordance with the procedure specified in ASTM Designation: E136-65. Temperature Specification: 1382 ± 10°F. Exposure period specification: After 5 minutes of exposure, test continued until thermocouples on specimens reached maxima.

TEST RESULTS

EXPOSURE TEMPERATURE, °F

3004 Alloy Specimen	Initial Temperature	Final Temperature	Observation
1	1390	1380	Noncombustible
2	1390	1380	Noncombustible
3	1390	1380	Noncombustible
4	1385	1380	Noncombustible
8112 Alloy Specimen			
1	1390	1380	Noncombustible
2	1385	1380	Noncombustible
3	1390	1380	Noncombustible
4	1390	1380	Noncombustible

Respectfully submitted,
SIGNET TESTING LABORATORIES, INC.

By Wilbur Chinn

Fig. A3.2 Signet Testing Laboratories, Inc. Noncombustibility Tests, Report dated September 30, 1968, confirming that aluminum alloys 3004 and 8112 were rated noncombustible in tests prescribed by ASTM Standard E136-65, "Standard Method of Test for Determining Noncombustibility of Elementary Materials"

Jul-25-00 01:06P TEMCOR 310 549 6807 P.05

LABORATORY CERTIFICATE

SIGNET TESTING LABORATORIES, INC.

TESTING AND INSPECTION OF CONSTRUCTIONAL AND INDUSTRIAL MATERIALS

ENGINEERS
CHEMISTS
METALLURGISTS

REPLY TO
1420 WEST WINTON AVENUE
HAYWARD, CALIFORNIA 94545
782-7919 AREA CODE 415

May 17, 1972

LAB NO.: 10263 P.O. No.: 395M909727

SAMPLES: Three (3) Samples of Aluminum Alloy

MARK: 3003 Alloy, 3105 Alloy and 5005 Alloy

DATE TESTED: May 5, 1972

REPORT TO: Kaiser Aluminum and Chemical Corporation
300 Lakeside Drive #7730B
Oakland, California 94604
Attention: R.A. Edgren

REPORT

TEST

The submitted samples of aluminum alloys were tested for noncombustibility property as prescribed by ASTM Designation: E136-65, "Standard Method of Test for Determining Noncombustibility of Elementary Materials".

TEST SPECIMENS AND PROCEDURES

The prepared test specimens were 1 1/2" wide by 1 1/2" thick by 2" long. The test was performed in accordance with the procedure specified in ASTM Designation: E136-65. Temperature Specification: 1382 + 10°F. Exposure period specification: after 5 minutes of exposure, test continued until thermocouples on specimens reached maxima.

TEST RESULTS	EXPOSURE TEMPERATURE, 0°F		
	Initial Temperature	Final Temperature	Observation
3003 Alloy Specimen			
1	1390	1365	Noncombustible
2	1390	1338	Noncombustible
3	1390	1365	Noncombustible
4	1390	1390	Noncombustible
3105 Alloy Specimen			
1	1390	1390	Noncombustible
2	1390	1385	Noncombustible
3	1390	1390	Noncombustible
4	1390	1390	Noncombustible
5005 Alloy Specimen			
1	1390	1390	Noncombustible
2	1390	1385	Noncombustible
3	1390	1385	Noncombustible
4	1390	1390	Noncombustible

Respectfully submitted,
SIGNET TESTING LABORATORIES, INC.

By [signature]

2cc: Kaiser Aluminum

Fig. A3.3 Signet Testing Laboratories, Inc. Noncombustibility Tests, Report dated May 17, 1972, confirming that aluminum alloys 3003, 3105, and 5005 were rated noncombustible in tests prescribed by ASTM Standard Method E136-65, "Standard Method of Test for Determining Noncombustibility of Elementary Materials"

WFRC

WARRINGTON FIRE RESEARCH CENTRE

Holmesfield Road, Warrington, Cheshire WA1 2DS.
Tel: (0925) 55116 Telex: 628743 WARRES G Telefax: (0925) 55419

Our Ref: 46083 - CWM/KC
17th August 1989

WIMPEY OFFSHORE
27 Hammersmith Grove
Hammersmith
London
W6 7EN

For the Attention of Mr. C. Barry

Dear Sirs

CONFIRMATION OF FIRE RESISTANCE TEST RESULTS

We confirm the results of a fire resistance test which was performed on you behalf and adopted the principles and performance criteria of the IM Resolution A517 (XIII) and the Exchange of Notes Agreement between th governments of the United Kingdom and the Nordic Countries.

The test construction consisted of a roof section taken from an accommodatio module. The assembly primarily consisted of a sandwich panel of two layers o aluminium sheet with a core of mineral fibre board, stated to be 150 kg/m nominal density. The mineral fibre board was adhesive fixed to the insides o the aluminium sheets. The thickness of the mineral fibre board was a nomina 83 mm and the aluminium sheets were nominally 1.6 mm thick.

The underside of the assembly exposed to the furnace was insulated by 50 mm o mineral fibre board stated to be of 110 kg/m³ nominal density, this wa adhesive fixed to the underside of the specimen.

The test construction incorporated a corner section of the roof and also joint in the sandwich panel.

The overall size of the test assembly was nominally 3000 mm long by 2400 m wide.

The results obtained from the test were as follows:

Insulation	: 110 minutes
Resistance to penetration of smoke and/or hot gases	: 110 minutes
Test discontinued	: 110 minutes

Cont. ....

Fig. A3.4 Warrington Fire Research Centre Report 46083-CWM/KC dated August 17, 1989, confirming that a properly insulated aluminum roof module resisted the penetration of smoke and provided the required insulation for the maximum length of the test, 110 minutes

Page 2

Mr. C. Barry
Wimpey Offshore

Our Ref: 46083 - CWM/KC
17th August 1989

DATE OF TEST : 15th August 1989

The test was witnessed by Mr. D. Collins a representative of Lloyds Register of Shipping.

Our full report covering this test will follow in due course.

Yours faithfully

C.W. MILES
Technical Officer
Structural Fire Protection
WARRINGTON FIRE RESEARCH CENTRE

KC(1248)

Fig. A3.4 (continued) Warrington Fire Research Centre Report 46083-CWM/KC dated August 17, 1989, confirming that a properly insulated aluminum roof module resisted the penetration of smoke and provided the required insulation for the maximum length of the test, 110 minutes

WFRC

WARRINGTON FIRE RESEARCH CENTRE

Holmesfield Road, Warrington, Cheshire WA1 2DS.
Tel: (0925) 55116 Telex: 628743 WARRES G Telefax: (0925) 55419

Our Ref: 48322 - MT/KC
1st February 1990

WIMPEY OFFSHORE
27 Hammersmith Grove
Hammersmith
London
W6 7EN

For the Attention of Mr. C. Berry

Dear Sir

FIRE RESISTANCE TEST RESULTS

We confirm the results of a fire resistance test which was performed on your behalf and adopted the principles and performance criteria of IMO Resolution A517 (XIII).

The test construction consisted of an aluminium faced bulkhead which was manufactured from sandwich panels and incorporated a single-swing/single-leaf doorset. The bulkhead had overall nominal dimensions of height 2995 mm and width 3010 mm and the doorset had overall nominal dimensions of height 2200 mm and width 1100 mm.

The bulkhead comprised four panels which consisted of a mineral fibre core nominally 83 mm thick and of 150 kg/m³ density which was faced using nominally 1.6 mm thick aluminium sheet. The aluminium sheets were glue fixed to the core material using an adhesive referenced 'X908F'.

The bulkhead incorporated an upper boom, lower boom and corner post which were insulated by the same method used to insulate the sandwich panels.

The doorset was bolt fixed to internal aluminium sections which formed the door frame housing. The leaf of the doorset had overall nominal dimensions of height 2034 mm, width 933 mm and thickness 75 mm and consisted of carbon steel door faces which encapsulated a ceramic fibre core material of 128 kg/m³ nominal density. The door leaf was hung within its frame on 1½ pairs of heavy duty hinges such that it opened away from the heating conditions of the test and was fitted with an 'Assa' triple action latch set and lever handles. The bulkhead was bolted into a refractory concrete lined steel restraint frame which was mounted to form the front vertical face of a gas fired furnace. The results were as follows:

Resistance to penetration of smoke and/or hot gases : 65 minutes
Thermal Insulation : 65 minutes

The test was discontinued after 65 minutes.

Cont. ....

Fig. A3.5 Warrington Fire Research Centre Report 48322-MT/KC dated February 1, 1990, confirming that a properly insulated aluminum bulkhead module resisted the penetration of smoke and provided the required insulation for the maximum length of the test, 65 minutes

Page 2

Mr. C. Berry
Wimpey Offshore

Our Ref: 48322 - MT/KC
1st February 1990

DATE OF TEST : 8th December 1989

The test was witnessed by Mr. F. Bayati a representative of Lloyds Register of Shipping.

Thank you for the detailed specification provided, our report will follow in due course.

Yours faithfully

M. THOMPSON
Technical Officer
Structural Fire Protection
WARRINGTON FIRE RESEARCH CENTRE

KC(1731)

Fig. A3.5 (continued) Warrington Fire Research Centre Report 48322-MT/KC dated February 1, 1990, confirming that a properly insulated aluminum bulkhead module resisted the penetration of smoke and provided the required insulation for the maximum length of the test, 65 minutes

CERTIFICATE OF INSPECTION & TESTS

carried out by

TRANSPORT

Marine Surveyors of the Department of ~~Trade~~

This is to Certify that

tests carried out by the Fire Insurers' Research and Testing Organisation (FIRTO), Melrose Avenue, Borehamwood, Hertfordshire WD6 2BJ show that aluminium alloy ships sides or deckhouse sides of suitable scantlings insulated with ROCKWOOL FIREBATTS 825 MINERAL WOOL INSULATION manufactured by ROCKWOOL LIMITED, PENCOED, BRIDGEND, MID GLAMORGAN, WALES, CF35 6NY are such as would be acceptable to this Department on United Kingdom registered ships to form 'A' Class divisions of A60 standard in compliance with the requirements of

a. The International Convention for the Safety of Life at Sea 1974, Chapter II-2, Part A, Regulation 3(c);

b. The Merchant Shipping (Passenger Ship Construction) Regulations 1980, Part I, Regulation 1(2);

c. The Merchant Shipping (Cargo Ship Construction and Survey) Regulations 1981, Part I, Regulation 1(2); and

d. The Fishing Vessels (Safety Provisions) Rules 1975, Part I, Rule 1(5);

subject to the following conditions:-

1. The ROCKWOOL FIREBATTS 825 mineral wool insulation is to have a density of 110 $\pm$ 10% kg/m^3 and is to be of the same composition as that recorded by the Department and as that of the insulation which was fitted to the bulkhead panel which was tested by the Fire Insurers' Research and Testing Organisation and referred to in their report FIRTO TE 4565.

2. The method of insulating the aluminium alloy ships side or deckhouse side is as follows:-

2.1 The insulation is to be at least 65mm in thickness over the plating and fitted in two layers of 40mm and 25mm minimum thicknesses, the first layer on the plating being

Fig. A3.6 Certificate of Inspection & Tests carried out by the Marine Surveyors of the London Department of Transport dated August 31, 1983, certifying that an aluminum deckhouse module properly insulated with Rockwool insulation met the requirements of the A60 (highest) classification for fire resistance

the 40mm minimum thickness. The insulation over the frames or stiffeners may be reduced to a minimum thickness of 40mm applied in a single layer.

2.2 The methods of attaching the insulation to the aluminium alloy ships sides or deckhouse sides and other relevant details must be as shown on the manufacturers drawings Nos 1044 to 1047 inclusive which were endorsed with the stamp of the Department on 20 July 1983 and no departure will be permitted therefrom without the prior consent of the Department.

2.3 Adjacent slabs of insulation are to be fitted tightly together and to the structure and the joints in the two layers are to be staggered.

2.4 The insulation is to be continued along the vertical and horizontal structure adjacent to the ships side or deckhouse side which is being insulated for a distance of not less than 450mm.

3. The containers in which the insulation is packed are to be clearly marked with the name of the insulation and its density.

4. The insulation is to prove satisfactory in service.

5. This acceptance is given on the understanding that this Department reserves the right to require check tests of the ROCKWOOL FIREBATTS 825 insulation to be carried out at any time and that ROCKWOOL LIMITED will accept full responsibility for informing shipbuilders or their sub-contractors of the approved method of fitting the insulation and the conditions of this acceptance.

I G Noble

Issued at the DEPARTMENT OF ~~TRADE~~ TRANSPORT, LONDON the 31 day of August 19 83

Fig. A3.6 (continued) Certificate of Inspection & Tests carried out by the Marine Surveyors of the London Department of Transport dated August 31, 1983, certifying that an aluminum deckhouse module properly insulated with Rockwool insulation met the requirements of the A60 (highest) classification for fire resistance

10 October 1983

'A' Class marine fire test

Bulkhead with Rockwool Firebatt 825, 80mm thick on both faces

Rockwool Ltd.

FIRTO

Marine fire test on an aluminium bulkhead insulated with 80mm-thick Rockwool Firebatt 825 on both faces

Technical Evaluation

Report

for Rockwool Ltd., Western Avenue, Industrial Estate, Bridgend, Mid-Glamorgan, CF31 3RT

Summary

A bulkhead consisting of an aluminium plate, 6mm-thick, strengthened by aluminium angle-section stiffeners and protected with 80mm-thick Rockwool Firebatt 825 on both faces was submitted for a fire test, with the stiffeners on the exposed face, according to the requirements for 'A' Class bulkheads of the Exchange of Notes and SOLAS (IMO) regulations on 10 October 1983. The bulkhead achieved the A60 classification for fire resisting divisions.

FIRE INSURERS' RESEARCH AND TESTING ORGANISATION
Melrose Avenue, Borehamwood, Hertfordshire, WD6 2BJ
Telephone: 01-207 2345 Telex: 291835

Fig. A3.7 Summary of Fire Insurers' Research & Testing Organization (FIRTO) Report dated October 10, 1983, confirming that an aluminum bulkhead properly insulated with Rockwool insulation achieved the A60 (highest) classification for fire resistance

Lloyd's Register of Shipping

71 Fenchurch Street, London, EC3M 4BS

Certificate No.: ICD/F83/697

Date: 21st December, 1983

ITEM: ALUMINIUM BULKHEAD - INSULATED WITH 80 mm ROCKWOOL FIREBATTS 825 ON BOTH FACES (40 mm OVER STIFFENERS)

MANUFACTURER: ROCKWOOL LTD.
PENCOED
BRIDGEND CF35 6NY

This is to certify that the above insulated Aluminium bulkhead, the construction and constituent materials, as detailed in the Fire Insurers' Research and Testing Organisation Fire Test Report FIRTO TE 4741 dated 16th November, 1983, will be accepted for compliance with this Society's Rules and Regulations and with the International Conventions for the Safety of Life at Sea 1960 and 1974, on ships classed with Lloyd's Register of Shipping and on ships for which the Society is responsible for the issue of the relevant certificate required by the above International Conventions as authorised by contracting Governments, as an:

A-60 CLASS BULKHEAD

NOTE: The foregoing certification is extended to cover the use of the materials in:

(i) Offshore Installations subject to the Society's Classification Rules;

(ii) Offshore Installations for which this Society is authorised to issue Certificates of Fitness, and similar licences, permits, etc.

This Certificate is valid for 5 years from the date of issue.

Deputy Chairman

J. T. Cullen
pro Secretary

Issued upon the terms of the Society's Regulations which provide that:

"In providing services information or advice neither the Society nor any of its servants or agents warrants the accuracy of any information or advice supplied. Except as set out herein neither the Society nor any of its servants or agents (on behalf of each of whom the Society has agreed this clause) shall be liable for any loss damage or expense whatever sustained by any person due to any act or omission or error of whatsoever nature and howsoever caused of the Society its servants or agents or due to any inaccuracy of whatsoever nature and howsoever caused in any information or advice given in any way whatsoever by or on behalf of the Society, even if held to amount to a breach of warranty. Nevertheless, if any person uses the Society's services or relies on any information or advice given by or on behalf of the Society and suffers loss damage or expense thereby which is proved to have been due to any negligent act omission or error of the Society its servants or agents or any negligent inaccuracy in information or advice given by or on behalf of the Society then the Society will pay compensation to such person for his proved loss up to but not exceeding the amount of the fee (if any) charged by the Society for that particular service information or advice, or that part thereof which caused the loss."

Form 1125 (11/78) (Rpt. 10c-Lon)

Fig. A3.8 Lloyd's Register of Shipping Certificate No. ICD/F83/697 certifying that an aluminum bulkhead properly insulated with Rockwool Firebatts is accepted for compliance with the International Conventions for the Safety of Life at Sea, and is accepted for use at sea

APPENDIX **4**

Fire Protection for Aluminum Alloy Structural Shapes

J.G. Kaufman and R.C. Kasser, Alcoa

Civil Engineering
March 1963

Adapted with the permission of the American Society of Civil Engineers, publishers of *Civil Engineering Magazine*

The increasing use of aluminum alloy structural members to take advantage of their light weight in building construction has resulted in a need for more information on the treatment of such members to provide fire protection equivalent to that required for steel members (Ref 1). Steel members are fire protected in such a way that the average temperature at any cross section is not expected to exceed 1,000 °F during a standard fire test (Ref 2, 3). Although the specifications are not specific as to the criteria used in establishing this limiting temperature, a study of the properties of low-carbon structural steel indicates that this limitation insures that steel members will:

- Retain yield strengths at least equal to the design allowable stresses, and hence have adequate load carrying capacity during the test exposure.
- Undergo no substantial change in their mechanical properties at room temperature as a result of the test exposure.

Limiting Temperatures

A study of the mechanical properties of aluminum alloys at elevated temperatures (Ref 4) indicated that the limiting temperatures necessary to meet these criteria are, as expected, lower than those for structural steel. The two criteria used were:

- To insure yield strengths at least equal to the design allowable stresses during the test exposure, the limiting temperature for aluminum would be 500 °F.
- To insure that there will be no substantial change in properties at room temperature as a result of the test exposure, the limiting temperature would be 375 °F.

Since it is not clear which of the two criteria were used in establishing the fire protection necessary for steel members, Alcoa has made preliminary tests of two columns to establish the amount of fire protection necessary to meet either criteria. The tests were made by the alternate procedure for structural steel columns described in ASTM E119-6l, "Methods of Fire Tests of Building Construction and Materials" (Ref 2).

Two columns tested

Two extruded 2014-T6 wide-flange columns, 9 ft long and 8 in. deep (8 WF 10.72: 8 × 8 × 0.288 in., 10.72 lb per ft) were used in these tests. They were mounted on Alclad 2014-T6 base plates, 1 × 24 × 24 in., and the assembly was in turn mounted on three legs consisting of 4 in. 2014- T6 I-beams. Portland-cement concrete caps 3 in. thick were cast around the top and bottom plates on each column.

Following assembly of each column, 16 thermocouples were attached to it. There were three each at four different levels as specified by ASTM E119-61 (Ref 2) (the top and bottom ones 2 ft from the end of the column and the others equally spaced in between), and four others near the top of the column.

The two columns were fireproofed by the common commercial procedure of coating with lightweight Vermiculite plaster, as shown in Fig. A4.1. The first column was wrapped with lath, and plaster was applied to a thickness of 2 in. from the face of the lath (including a scratch coat ¼ in. thick, the base coat and a finish coat of white lime plaster 1/16 to ⅛ in. thick). The second column was wrapped with lath and plaster was applied to a thickness of about 1 ½ in. (including a layer of scratch coat ¼ in. thick and the base coat); then a second layer of lath was wrapped around it and additional plaster was applied so that the total thickness was 3 in. from the face of the lath (including a finish coat of white lime plaster 1/16 to ⅛ in. thick).

With both columns, standard 3/8-in, 3.4-lb diamond mesh was used. Keystone key corner beads were employed to insure the indicated thickness of plaster on the face of the lath. The plasters used were:

- Scratch coat—one 100-lb bag of glass-fibered gypsum to 2 cu ft of Vermiculite aggregate
- Regular coat—one 100-lb bag of glass-fibered gypsum to 2 ½ cu ft of Vermiculite aggregate
- White finish coat—one 100-lb bag of unfibered gypsum to one 100-lb bag of lime

Bestwall glass-fibered gypsum (ASTM C-35) and Zonolite Vermiculite aggregate (ASTM C-22) were used. The fire protection was done by McNulty Brothers of Chicago, a contractor experienced in the application of fire-protection systems. The completed columns were allowed to set for four months before testing. Figure A4.2 shows one of the fire-protected columns before testing.

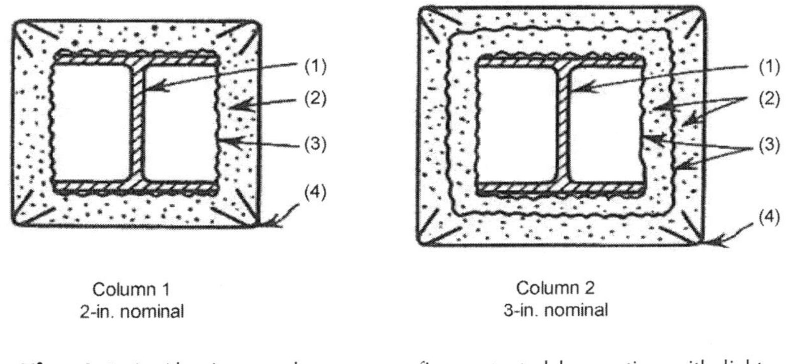

Fig. A4.1 Aluminum columns were fire protected by coating with lightweight Vermiculite plaster. Numbers designate materials as follows: (1) 8 WF 10.72 column, (2) Vermiculite plaster, (3) lath, (4) Keystone key corner beads

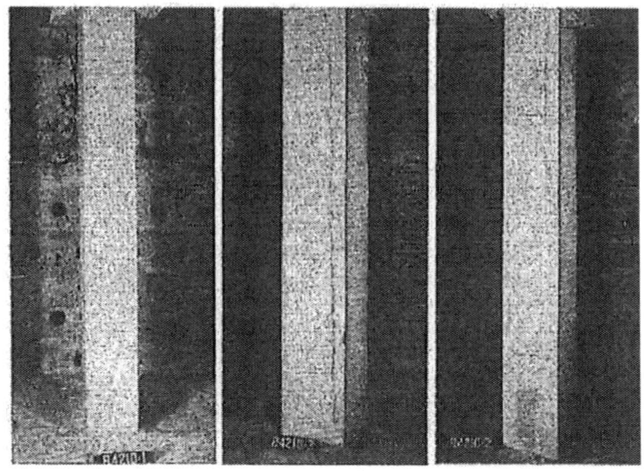

Fig. A4.2 Fire-protected column ready for test (left); fire-protected columns after testing (center and right)

Test procedure and results

The two columns were tested individually by being placed in a gas-fired furnace, 9 × 9 × 9 ft, and subjected to the standard ASTM time-temperature exposure (Ref 2, 3). The temperatures of each column during the test, as indicated by the thermocouples, were recorded on Leeds and Northrup continuous-recording potentiometers.

In the test of the column covered with 2 in. of lightweight Vermiculite plaster, the calcining period (the time for the water in the plaster to dry out) was approximately 65 minutes, after which the temperature of the column gradually increased at an increasing rate to a maximum of 8 °F per min. The average temperature of the column on the hottest level (as defined by a group of thermocouples) exceeded 375 °F in 2 hours 13 min and exceeded 500 °F in 2 hours 29 min from the start of the test.

In the test of the column covered with 3 in. of plaster, the calcining period was about 2 hours, and the maximum rate of increase of column temperature was about 5 °F per min. The average temperature at the hottest level exceeded 375 °F after 4 hours 7 min, and 500 °F after 4 hours 30 min. The appearance of the columns after testing is shown in Fig. A4.2.

The average temperatures in the hottest region of the columns are summarized in Fig. A4.3 and a cross plot showing hours of protection as a function of plaster thickness is given in Fig. A4.4. From Fig. A4.4, the thicknesses of plaster required for the protection of aluminum alloy members for various periods were estimated, and are shown in Table A4.1 along with the corresponding thicknesses for steel members.

Preliminary tests of floor panels have indicated that the same thicknesses shown in the table can be used to provide adequate fire protection for aluminum floors and roofs.

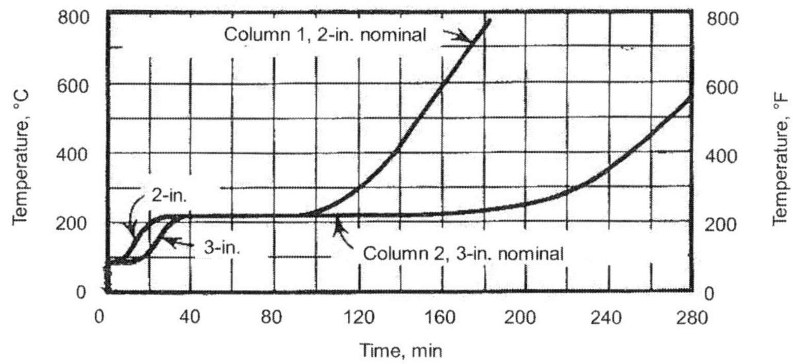

Fig. A4.3 Curves show average temperature at hottest areas within columns during fire test exposure.

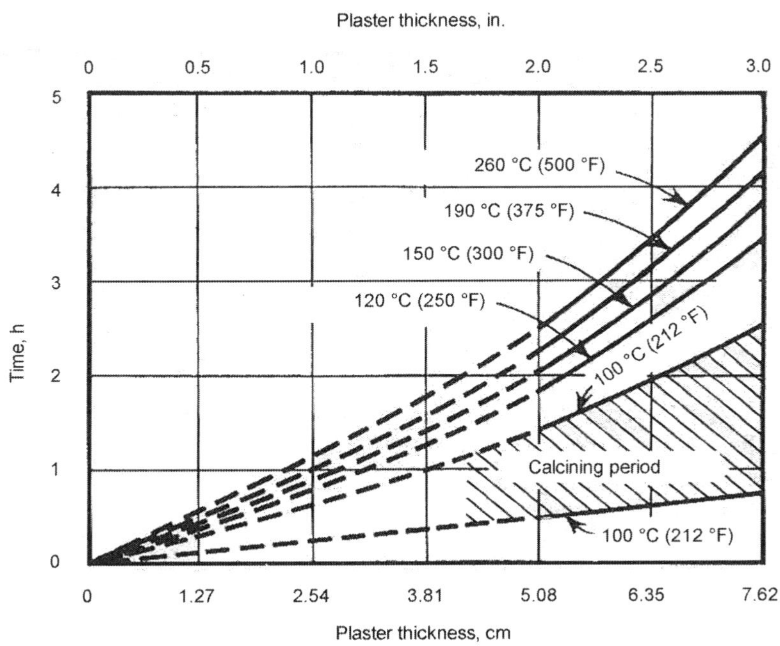

Fig. A4.4 Hours of protection are shown as function of plaster thickness.

Table A4.1 Plaster thicknesses for fire protection of aluminum alloy columns

Fire protection period, hours	Plaster thickness (in.)(a) to prevent temperature of Al columns in excess of: 375 °F(b)	500 °F(c)	Plaster thickness (in.) required for steel columns (1000 °F)
1	1 ¼	1 ⅛	3/4
2	1 7/8	1 3/4	1
3	2 ½	2 3/8	1 3/8
4	3	2 7/8	1 3/4

(a) From the face of the lath; (b) To insure no substantial change in properties at room temperature as a result of test exposure; (c) To insure yield strengths at least equal to the design allowable stresses during the test exposure

The data for the aluminum columns may be reduced to an algebraic form (Ref 5) to permit calculations of fire retardance periods for a wide range of column size and plaster coatings. The basic form of the equation is:

$$R = c\left(\frac{D^2 - ad^2}{D}\right)^n$$

where

R = fire retardance period, hours
D, d = outside and inside dimensions of concrete, in.
n = constant equal to 1. 7
c, a = constants dependent on materials

From the data for the aluminum columns, constants c and a can be determined (c = 0.1, a = 1.0), so that the fire retardance periods for aluminum columns can be computed by:

$$R = 0.1\left(\frac{D^2 - d^2}{D}\right)^{1.7}$$

From these data, it is clear that aluminum alloy structural members can be fire protected by the practical and commercially acceptable procedure of coating with lightweight Vermiculite plaster. This is the same procedure used to provide fire protection for steel members, except that a greater thickness of plaster is required for the aluminum alloys.

BIBLIOGRAPHY

1. "Fire Protection Equipment List," Underwriter's Laboratories, 1962.
2. ASTM E 119-61. "Methods of Fire Tests of Building Construction and Material," ASTM Standards, 1961, pt 5, p. 1136.
3. "Standard for Fire Tests of Building Construction and Materials," Underwriter's Laboratories, UL 263. 1 Jan. 1955.
4. Military Handbook 5, "Strength of Metal Aircraft Elements," March 1961.
5. Appendix 2 of "Fire Retardance Classification of Building Constructions," Building Materials and Structures Report BMS92, National Bureau of Standards, Oct. 7, 1942.

APPENDIX 5

ALFED Fact Sheet 3
Alumium and Fire

Reprinted with the permission of the Aluminium Federation, Ltd
Birmingham, United Kingdom
July, 2004

Introduction

Aluminum and its alloys are the most commonly used nonferrous metal materials and they find wide application in transport, building, packaging, general and electrical engineering. The excellent physical and mechanical properties of aluminum alloys lead to their widespread use. The alloys are light but strong and their use in transport saves fuel over the whole of the vehicle lifetime. The alloys can be fabricated or cast by all conventional processes, joined by welding or adhesive bonding and given a variety of surface coatings. These coatings, together with the inherent high degree of resistance to corrosion by the alloys, allow the use of aluminum components in aggressive environments. Finally, after use even over many years, the aluminum can be recycled, with a saving of energy of 95% of that required through the primary smelter production route.

Because of the widespread use of aluminum alloys in building, transport, home appliances and offshore structures, it is necessary to address the issue of aluminum and fire and to answer the question, "Does aluminum burn?"

The answer is, of course, "No". Each year hundreds of thousands of tonnes of aluminum scrap are fed into remelt furnaces and heated up to and beyond the melting point. Aluminum melts when the temperature exceeds the melting point, it does not burn. If it did, the recycling of aluminum would not be possible.

During and following the Falklands conflict between Great Britain and Argentina, several misleading statements appeared in the press, suggesting that aluminum alloys, used in the superstructure of some of the ships that were sunk, had burned and contributed to the loss of these ships. Of the nine ships sunk in this conflict, only three had aluminum superstructures (Fig. A5.1). All three vessels had steel hulls and in each case the damage inflicted suggested that these vessels would have sunk regardless of the materials of the superstructure. In no case did aluminum burn. The first British destroyer to be sunk, HMS Sheffield, was widely reported to have an aluminum superstructure. This was, in fact, an all-steel ship with a steel hull and a steel superstructure. The Defence White Paper published on December 14, 1982 concluded, "There is no evidence that aluminum has contributed to the loss of any vessel." Similar conclusions were reached by the Ministry of Defence Working Party convened to review ship design. The Chairman of that Working Party was reported in the Financial Times, December 24, 1982, "I am not aware of any evidence to suggest that any ship was lost because of the use of aluminum in its construction, nor was there any evidence that aluminum or aluminum alloys had burned or suffered from a series of small explosions. Aluminum, like any other material, has advantages and disadvantages in any given set of circumstances. Where the balance is in favor, aluminum should be used in warships or elsewhere."

In fire tests on aluminum materials, when the temperature exceeds the melting point, in the range 600 to 660 °C (1110 to 1220 °F), the aluminum surface exposed to the fire can be seen to melt, but it does not burn. At the end of the fire test, the metal remains as a resolidified pool.

Fig. A5.1 The superstructures of some warships are made of aluminum

Aluminum in the form of finely divided powder or flake oxidizes exothermically, but this is a very special case because of the very large surface-area-to-weight ratio. Aluminum then behaves in a similar way to other finely divided materials such as iron and titanium, tea, flour and coal, all of which will also readily oxidize exothermically in the powder form.

Aluminum in a Fire

If aluminum is involved in a fire and the temperature rises to above its melting point, the metal begins to melt. The melting point of a metal, like its density, strength, or corrosion resistance, is a characteristic which can be measured and used to design the most effective component for any particular application.

The thermal conductivity of aluminum is around four times that of steel and its specific heat twice that of steel. This means that heat is conducted away faster and a greater heat input is necessary to bring the same mass of aluminum to a given temperature, compared with steel. Where an aluminum structure is exposed to the heat of a fire, the relatively high thermal conductivity enables the heat to be rapidly conducted away from the exposed area. This helps to reduce hot spots where significant localized property loss could occur, so extending the serviceability period. It will, however, cause the temperature to rise elsewhere. The extent of dissipation of heat elsewhere in the structure will depend on the degree of thermal insulation provided to the aluminum elsewhere in the structure, necessary to provide fire protection in that area. The high reflectivity of weathered aluminum is 80 to 90 percent compared with 5% for painted steel and 25% for stainless steel. This is of considerable benefit and will assist in prolonging endurance of an aluminum structure in a fire.

Figure A5.2 is a good example of the behavior of aluminum in the massive form in a fire. A car, with aluminum alloy wheels, was caught in a forest fire that swept over the car and moved on. Afterwards it was found that the aluminum wheels had melted, molten aluminum had run off and collected in a pool of metal which solidified as the fire moved on and the temperature fell. The aluminum had not burnt.

Aluminum in Building

Aluminum alloy components are widely used in building as cladding and roofing materials, windows and doors. As defined by BS 476: Part 4 and the 1974 SOLAS Convention (as amended) aluminum alloys are "noncombustible" and also provide Class 1 surface spread of flame to BS 476: Part 5. In addition, BS 476: Part 3 covers external fire exposure roof tests and the classifications laid down in the standard range from AA to

Fig. A5.2 Wheels on fire: aluminum wheels melted by the inferno that ripped through San Bernadino, California, USA in October 2003

DD. The first letter refers to the fire penetration performance and the second letter to the surface spread of flame. Aluminum and its alloys are rated AA, the highest possible under this classification system. Materials are also tested for fire propagation performance to BS 476: Part 6 (1989) and coating systems are taken into account. Aluminum achieves excellent ratings under this Standard. Three principal methods of fire protection are employed, using fire resistant insulating layers protecting an underlying aluminum component. Examples have used ceramic fiber, intumescent coatings applied to the aluminum element or composite systems with aluminum external skins. The latter have been demonstrated to be applicable both as load-bearing elements and as add-on panel systems. The aluminum skin on the exposed fire side of the system is sacrificial and melts, revealing a supported fire insulation material which provides the required period of fire performance and protects the remaining aluminum elements of the system. A small increase in insulation thickness is usually required to take account of the lower maximum working temperature limit of aluminum compared to steel. The use of radiation shielding around structures such as stair towers and walkways can provide protection, not only to personnel during a fire but prevent the temperature of the aluminum structure exceeding the working limits during the design time period.

As with all metallic materials, as the temperature increases the strength of aluminum alloys is reduced at a rate dependent on the alloy. The structural aluminum alloys have useful maximum working temperature limits that range from 200 to 250 °C (390 to 480 °F). Above this temperature the strength is significantly reduced.

Aluminum in Marine and Offshore Applications

Aluminum alloys, particularly the aluminum-magnesium alloys, have been used since the 1930s in many marine applications. In offshore structures, aluminum alloys are used in many applications (Fig. A5.3) such as helicopter landing decks, stairs and walkways, link bridges, housings for electrical equipment and even complete living quarter modules. Considerable cost savings can accrue by the use of aluminum without any increased fire risk. The use of aluminum also keeps the total weight down and keeps the center of gravity as low as possible.

Fig. A5.3 Aluminum is used in many off-shore applications, including helidecks and accommodation modules on oil rigs

Index

A

B

C

D

E

P

R

S

T

U

V

W

Z